PLANETS

A GUIDE TO THE SOLAR SYSTEM

a Golden Guide® from St. Martin's Press

by
MARK R. CHARTRAND, Ph.D.

Illustrated by
RON MILLER

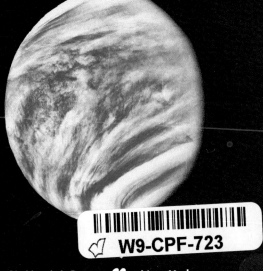

St. Martin's Press　ℳ　New York

FOREWORD

Welcome to the solar system! Over the decades of the space age, we have learned more about our celestial neighbors than in any other period in history.

This book is for everyone who wants to explore these worlds. It provides information about them and tells you where and when they can be seen in the sky. This book, a companion to the Golden Guide *Stars*, is both an armchair guide and a portable field guide.

The author would like to express his thanks to many people, but particularly to the following: Caroline Greenberg and Remo Cosentino at Golden Press, who initiated the book and tirelessly gave their expertise, encouragement, and friendship; Bonny Lee Michaelson for help and encouragement; colleagues of the National Space Society, especially Leonard David; and Ron Miller for his fine illustrations. A special thanks is also owed to the scientists, managers, and associated investigators of the National Aeronautics and Space Administration for achieving the spectacularly successful planetary exploration missions that made possible many of the photographs in this book. And acknowledgment is due to the American public for making these explorations possible through their support.

PLANETS. Copyright © 1990 by St. Martin's Press. Illustrations copyright © Ron Miller. All rights reserved. Printed in the United States of America. No part of this book may be used or reproduced in any manner whatsoever without written permission except in the case of brief quotations embodied in critical articles or reviews. For information, address St. Martin's Press, 175 Fifth Avenue, New York, N.Y. 10010. www.stmartins.com

A Golden Guide® is a registered trademark of Golden Books Publishing Company, Inc., used under license.

ISBN 1-58238-146-1

CONTENTS

Ancient skywatchers

INTRODUCTION

The planets and their satellites, asteroids (the minor planets), and the Sun make up the solar system, our cosmic neighborhood. If we could travel at the speed of light, 186,000 miles a second, it would take only eight minutes to reach the Sun, and only half a day to cross the solar system from one side of Pluto's orbit to the other. The nearest stars, by comparison, are several light-years away.

MYTHOLOGY is the way early peoples tried to explain what they saw in the sky. They could easily see the Sun and Moon, five bright planets, occasional comets, and frequent meteors, but they did not know what they were. To them, everything in the sky was magical, some sort of god or demon. Astrology, the false claim that events in the heavens control our lives and predict the future, grew out of these myths.

The ancients tried to find patterns in the sky and the movement of the objects in it. Among the fixed stars they saw "pictures"—the constellations. They noticed that the Sun, Moon, and planets wandered among the stars. Many ancient peoples built monumental structures—Stonehenge is an example—to observe the sky.

ASTRONOMY is the science of observing the universe, using logical procedures and continual testing of theories. As we refine our knowledge, some of the facts and numbers in this book will have to be revised. Astronomy is an international endeavor. You see the same sky the ancients saw, but you have the benefit of much more knowledge to explain what you observe. Modern astronomy is only a few centuries old.

SKYWATCHING is an enjoyable pastime. Field guides, such as *Stars* and *Skyguide* (see the bibliography, p. 157), can help you identify the stars and constellations, while astronomy magazines keep you in touch both with events in the sky and with recent discoveries of astronomy and the space program.

Binoculars, preferably 7 × 50 size, are very useful for scanning the sky. They are also comparatively inexpensive. Later you may wish to step up to a telescope. Before you buy one, consult a good guide for advice and avoid poorly made department-store telescopes.

PLANETARIUMS are great places to become familiar with the sky. Often planetariums also offer public observing through telescopes, and courses in astronomy. Amateur astronomical societies have frequent meetings and "star parties" where you can look at the planets through small telescopes. Professional observatories are usually too busy to allow the public to use the instruments at night.

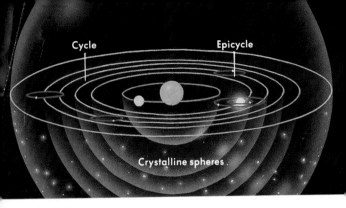

Cycle

Epicycle

Crystalline spheres

The Ptolemaic system

EARLY THEORIES OF THE PLANETS

The earliest skywatchers who kept records were the Sumerians, Babylonians, and Egyptians. They were followed by the Greeks, who continued to try to find order in the sky and its motions. Everything in the sky was thought to be some sort of star: fixed stars that didn't change and didn't move relative to one another, shooting stars (meteors) that flashed across the sky, hairy stars (comets) that appeared occasionally. They gave the name *planetes,* meaning "wanderers," to the Sun, Moon, and planets that move against the background of fixed stars.

TO THE GREEKS, the Earth was the center of the universe, and everything moved around it. In their philosophy, everything above the Earth was celestial and perfect. The perfect shape was the sphere, so they thought that the planets (including Sun and Moon) were attached to huge turning transparent crystalline spheres centered on the Earth, causing planetary motions.

THE PTOLEMAIC SYSTEM was set forth by Claudius Ptolemy, a Greek astronomer working in Alexandria, Egypt, in the 2nd century A.D. He proposed that planets move on little circles, called epicycles, while the center of each epicycle moves in a circular path, a "cycle" or "deferent," around the Earth. By choosing the proper sizes and speeds for the cycles and epicycles, this theory could account for irregularities in the observed motions of the planets. Ptolemy's scheme was thought to be true for almost 1,400 years.

THE TYCHONIC SYSTEM was proposed by Tycho Brahe, a Danish astronomer, in the late 16th century. He thought the Moon and Sun revolved around the Earth, but that all the planets revolved around the Sun. Tycho was not right, but his careful naked-eye observations led Johannes Kepler to figure out just how they *do* move (see p. 31).

PRE-TELESCOPIC INSTRUMENTS consisted of mechanical devices for measuring *angles,* and hence positions, of things in the sky. Sundials and hourglasses measured the passage of time, and accurate clocks did not come along until the 17th and 18th centuries. The most common astronomical observing instruments were astrolabes, which were often beautiful works of craftsmanship as well as practical tools. Tycho and others also used large quadrants, almost the size of a room, to measure angles in the sky very accurately.

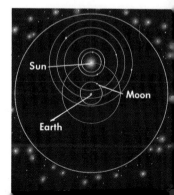

The Tychonic system

Copernicus' Sun-centered system

THE COPERNICAN REVOLUTION

The Polish-German astronomer Nicolaus Copernicus, who died in 1543, revolutionized our ideas about the universe when he proposed, correctly, that the Sun is the center of the solar system, and the Earth is just another planet. His controversial ideas eventually prevailed.

Several decades later, Tycho disproved two of the key beliefs the ancients had about the sky, and so helped prove Copernicus right. One ancient belief was that the fixed stars were unchanging, but in 1572 Tycho discovered a new star, called a *nova*. (We know now that novas are actually older stars blowing up and getting temporarily brighter.) Tycho also disproved a belief that comets were really hot gasses high in Earth's atmosphere. Careful measurements of a comet showed it was beyond the Moon. One of Tycho's assistants, Johannes Kepler, used Tycho's observations to deduce the laws by which planets move around the Sun (see p. 31).

THE TELESCOPE was adapted for astronomical use by Galileo Galilei, in Florence, about 1610. With it, distant objects were brought nearer and clearer, fainter objects became visible, and measurements were more precise. Among Galileo's first discoveries were four satellites revolving about Jupiter. This discovery also helped disprove the older ideas. For these contributions, which upset many conservative beliefs, Galileo was arrested and imprisoned by the Catholic Church until the end of his life, and his books were long banned.

Over the next few decades, Galileo and others, such as Cassini, Roemer, and Huygens, discovered many new things: the rings of Saturn, more satellites, sunspots, and many of the phenomena we know today. With them began the modern age of astronomy.

Galileo Galilei and his new telescope

THE SOLAR SYSTEM IN THE UNIVERSE

The solar system is only a very, very small part of the universe. We do not know how big the universe is, but astronomers can see several billions of light-years away with their largest radio and optical telescopes. (A *light-year*, the distance light can travel in a year, is approximately 5.8 trillion miles.) The universe is between 10 and 20 billion years old. Our solar system and Sun are only about 4.5 billion years old.

THE LARGEST STRUCTURES we know of are clusters of galaxies, islands of stars, tens or hundreds of millions of light-years across, containing from a few dozen to thousands of galaxies.

GALAXIES are of different shapes and sizes. Some are flat disks called spiral galaxies (our own Milky Way galaxy is a spiral). Others are globular or ellipsoidal, while still others are irregular. Galaxies contain from a few billion to a trillion stars and range in size from a few tens of thousands of light-years across to several hundred thousand light-years in diameter. Galaxies are typically spaced a few million light-years apart.

STARS, together with clouds of dust and gas (called *nebulas*), typically are spaced a few light-years apart. Stars are huge gaseous balls, generating energy through nuclear fusion in their cores. The Sun is an average-sized star. Other ordinary stars range in size from about a tenth the size of the Sun to perhaps ten times its size. Extraordinarily small stars, such as neutron stars, may be only a few miles across, the size of a large city, while the largest stars are billions of miles in diameter, the size of the orbit of Jupiter or Saturn in our solar system.

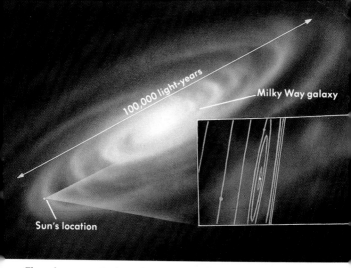

The solar system in the galaxy

PLANETS are cold, relatively small bodies (compared to most stars) circling the Sun (and perhaps other stars). The diameter of the solar system, measured across the orbit of the outermost planet, Pluto, is only about 11 light-hours.

SATELLITES, loosely called "moons," are smaller bodies orbiting the planets. Some of them may be as large as the smaller planets. *Asteroids* (minor planets) are small bodies orbiting the Sun.

NAMES of celestial bodies such as newly found asteroids and satellites, and names for such features on planets and satellites as mountains and craters, are proposed by their discoverers and must be approved by the International Astronomical Union (IAU). Comets are named for their discoverer(s). Smaller bodies may not have names, only catalog numbers.

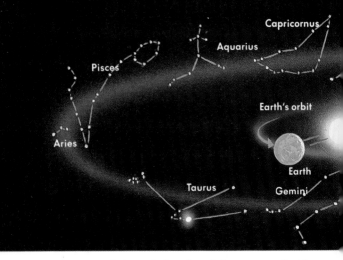

The Sun, Moon, and planets look as though they move against the background of the constellations of the zodiac.

OBSERVING THE SKY

THE ZODIAC

The zodiac is the region of the sky through which the Sun, Moon, and planets pass, as seen from Earth. It is a band about 30 degrees north and south of the *celestial equator*, the projection of the Earth's equator into the sky. For simplicity, we pretend that all the stars are attached to a great "celestial sphere" some large, indefinite distance away. These stars, and the patterns of the stars called constellations, are the background against which the planets move.

Following the practice of the ancients, the band of the zodiac is divided into 12 constellations, most of which were thought to represent living creatures. "Zodiac" means "circle of the animals" in Greek. The 12 constellations of the

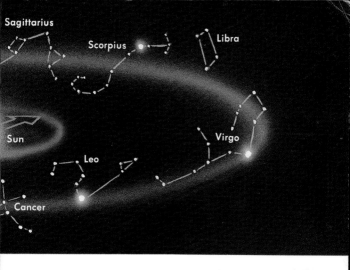

zodiac are: Aries, Taurus, Gemini, Cancer, Leo, Virgo, Libra, Scorpius, Sagittarius, Capricornus, Aquarius, and Pisces.

A convenient way of locating the planets is to give the name of the constellation of the zodiac they are "in," which really means "in front of." Because of changes in the sky over the past 2,000 years, which altered the orientation of the constellations, today the planets actually travel through more than the 12 classical constellations.

As seen from the Earth, the Sun seems to travel during a year around the zodiac, always along a line or path in the sky called the *ecliptic*, the plane of the Earth's orbit around the Sun projected out onto the celestial sphere. Because the Earth is tilted over 23 degrees, the ecliptic and the celestial equator are inclined by the same amount. The orbits of the Moon and planets are tilted slightly to the ecliptic, so their paths in the sky do not lie exactly along the ecliptic, but they are always close—except for Pluto.

PLANETARY PHENOMENA refers to the appearance in the sky and the relative positions of planets with respect to the Sun or each other. The angle between the Sun and a planet, as we see them from Earth, is a planet's *elongation*.

"EVENING STAR" is a term given to any planet that remains in the sky after the Sun has set. Such a planet will be east of the Sun—that is, it will have an easterly elongation.

"MORNING STAR" refers to a planet that is in the sky at dawn. It will be west of the Sun, and so have a westerly elongation. All the planets are at times morning stars and at other times evening stars.

INFERIOR PLANETS—Mercury and Venus—are those with orbits closer to the Sun than Earth's orbit. They are never seen at a great angle from the Sun. When an inferior planet is aligned with the Sun, it is said to be in *conjunction*. *Inferior conjunction* ("I" on diagram below) occurs when the planet is on the near side of the Sun, *superior conjunction* ("S") when it is on the far side of the Sun. The greatest angular distance between an inferior planet and the Sun occurs when our line of sight to the planet is tangent to the planet's orbit, called *greatest eastern elongation* ("E") when the planet is east of the Sun and *greatest western elongation* ("W") when it is west of the Sun.

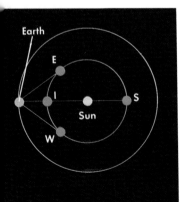

Inferior planet phenomena

Inferior planets show *phases*, like the Moon, related to where they are in their orbits. They are at *full phase* at the time of superior conjunction, *half phase* at times of greatest elongation, and *new phase* (hence invisible) at the time of inferior conjunction.

SUPERIOR PLANETS—Mars, Jupiter, Saturn, Uranus, Neptune, and Pluto—are those with orbits larger than Earth's, and they may have any elongation. When on the far side of the Sun, as seen from Earth, they are said to be in *conjunction* ("C" on diagram below). When opposite the Sun, they are in *opposition* ("O"). When they are at right angles to the Sun, they are said to be in *quadrature* ("Q"). Such planets have only a narrow range of phases, being almost fully illuminated as seen from Earth at all times.

THE SYNODIC PERIOD of an object is the length of time it takes for the object to return to a previous relative position, such as opposition or conjunction.

VIEWING THE PLANETS For the inferior planets, times of greatest elongation are best for viewing. The superior planets are brightest and visible longest at night around the time of opposition—when the Earth is directly between the planet and the Sun.

Planets are not visible close to the times of conjunctions, for then they are too close to the Sun in the sky.

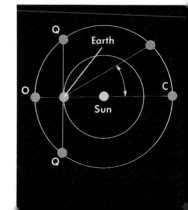

Superior planet phenomena

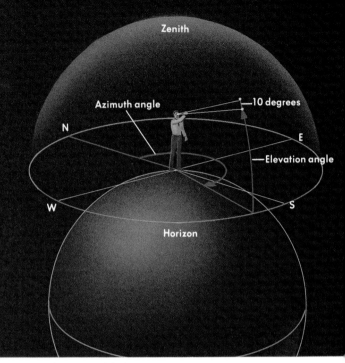

An observer and the horizon

FINDING THINGS IN THE SKY

What you see depends upon when you look. Sometimes the things you want to see are up in the daytime. The Moon is bright enough to be seen in the daytime, but the stars and planets are overpowered by the brightness of the daytime sky. As the Earth turns, the sky seems to revolve about us, moving west 15 degrees every hour.

To locate things in the sky, you must first know which direction is north. You can use a magnetic compass, or, at night, look for the North Star. For those in the northern hemisphere, the celestial equator will start at the horizon in the east, rise upward and to the right toward the south, reaching its highest point due south, and then decline downward to the right, intersecting the horizon due west, opposite east. All planetary objects will appear within 30 degrees on either side of this invisible celestial path.

AZIMUTH is a measure of direction around the horizon starting at 0 degrees due north, moving clockwise through 90 degrees at east, 180 degrees at south, 270 degrees at west, and 360, or 0, degrees again at north.

ELEVATION of an object is its angle above the horizon, from 0 degrees at the horizon to 90 degrees directly overhead (a point called the *zenith*). A useful rule of thumb is that your fist, held out at arm's length, is about 10 degrees across.

FINDING CHARTS Each planet in this book has a "finding chart" that shows the planet's path through the zodiac during upcoming years or in the morning and evening skies. (The chart for each planet can be found in the section on "Observing.") Tables give the best times for viewing each planet. For Mercury, this would be the week or so around the dates of greatest elongations; for Venus, several weeks around the greatest elongations; for the other planets, the several months around the times of opposition.

Sky motion around Earth

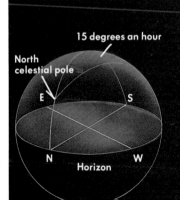

15 degrees an hour

North celestial pole

The atmosphere distorts light rays

EFFECTS OF THE ATMOSPHERE on what we see in the sky are due to our looking through miles of turbulent air. When we look at objects near the horizon the effects are greater, because we are then looking through more air.

TWINKLING, or scintillation, is caused by air layers slightly bending the light as it passes toward us. Stars, which are so far away they appear as mere points of light, twinkle more than planets do. Planets, which are closer to us, appear as small disks of light; they usually twinkle less. For objects close to the horizon, the air may act like a prism, causing a star or a planet to display flashing colors.

REDDENING occurs because the atmosphere is not equally transparent to all colors of light. As a beam of light travels, the air removes some of its blue light, scattering it in all directions and causing the sky to appear blue. The remaining light reaching your eye looks redder. The more air a light beam travels through, the greater the reddening, which is why the Sun and Moon appear reddish when they are low in the sky. Air pollution may accentuate this effect.

REFRACTION is a bending of the light from an object in the sky, making the object appear slightly higher in the sky than it is. The effect is greatest closer to the horizon, where it seems to flatten the rising or setting Sun and Moon.

18

THE MOON ILLUSION

The moon illusion is the name given to the optical illusion that the Moon (or Sun) appears larger when it is near the horizon than when it is high in the sky.

This seems to be caused by the fact that, when near the horizon, the Moon is seen near familiar foreground objects, such as trees or buildings or the horizon itself. Because the brain cannot determine how far away the Moon is, it incorrectly interprets what we see. The effect seems most noticeable for the Moon when it is near full phase.

One way to prove that the Moon does not change size is to photograph it rising, and then again, with the same camera and lens, when it is high in the sky. The image of the Moon on the slide or negative will measure the same linear size.

Full moon rising   CARNEGIE

Apollo 11 astronauts on the Moon (1969)   NASA

SOLAR SYSTEM EXPLORATION

Until the space age, we could only study the universe from Earth with instruments that had to look upward through miles of turbulent, filtering, polluted atmosphere. Now we send our robot observers throughout the solar system.

THE SUN has been examined by several special space observatories near Earth, including the Orbiting Solar Observatory and the Solar Maximum Mission. The distant effects of the solar wind have also been studied by probes called IMP, for Interplanetary Monitoring Platform, and by sensors on space probes heading to other planets. One of the IMP craft was later used to study a comet.

Voyager passing Saturn (1977) NASA

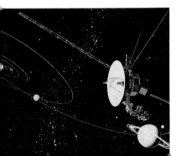

MERCURY has been explored by only one space probe as of 1989, Mariner 10, in 1973 and 1974. Its flight allowed it to pass this innermost planet three times, mapping the surface.

VENUS has been extensively investigated. The U.S. probe Mariner 2 flew by in 1962, revolutionizing our knowledge of this cloud-wrapped world. Since then, other American visits have been made by Mariner 5 in 1967, Mariner 10 in 1974, and Pioneer Venus, which in 1978 orbited the planet and sent a probe into its atmosphere. Most recently the Magellan spacecraft made detailed maps of the surface. By 1989, the Soviet Union had sent 13 spacecraft, all named Venera, to Venus. Some have even landed on the surface and taken a few photographs before the hostile atmosphere of that planet destroyed the probes. These are the only surface photographs we have of any planetary body besides the Moon and Mars.

EARTH, too, has been explored with spacecraft. Meteorological satellites monitor our atmosphere. Closer-orbiting Landsat earth resources satellites have also greatly improved our knowledge of our planet and its biological, geological, and hydrological systems.

Hubble Space Telescope   NASA

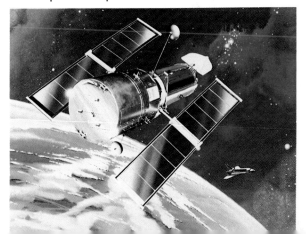

THE MOON, the only planetary body besides Earth explored by humans in person, was visited during the American Apollo program from 1969-72. More than 800 pounds of moon rocks were brought back for study. Before that, several U.S. and Soviet probes orbited or landed on the Moon to pave the way. The U.S.S.R. has sent robot landers to gather and bring back rock samples from the surface.

From Viking Lander 2, showing frost on Mars (1977)   NASA

MARS was first photographed close up by the American Mariner 4 in 1965. Three other Mariner craft and several Soviet spacecraft have visited the Red Planet, although the latter were not successful. The climax of Martian investigation came in 1976 when two U.S. Viking spacecraft orbited the planet and sent landing craft down to the surface. These landers operated for years, sending back photographs, soil analyses, and meteorological data.

JUPITER saw the flybys of Pioneer 10 in 1973, Pioneer 11 in 1974, and Voyagers 1 and 2 in 1979. They discovered new satellites and a ring, and photographed and measured details in Jupiter's atmosphere invisible from Earth. The computer-enhanced photographs of Jupiter are some of the most beautiful ever taken.

SATURN was visited by Pioneer 11 in 1979, by Voyager 1 in 1980, and then by Voyager 2 in 1981, after their swings by Jupiter, whose gravity helped get them out to the Ringed Planet. They discovered new moons and myriad rings.

URANUS was visited by Voyager 2 in 1986, after a gravity assist from Saturn. It discovered many new satellites, new rings, and a puzzling tilted magnetic field that has yet to be fully explained.

NEPTUNE was visited by Voyager 2 in August, 1989, the last Voyager encounter before leaving the solar system.

Landsat 4 photograph of Earth (1982) NASA

TWO COMETS have been probed. The first was Comet Giacobini-Zinner in 1985, reusing one of the IMP spacecraft, renamed ICE (International Cometary Explorer). Data from ICE helped space scientists from Europe, Japan, and the Soviet Union send probes to Comet Halley.

STARS, NEBULAS, AND GALAXIES as well as planets have been observed by the Orbiting Astronomical Observatory from near-Earth orbit, and by other specialized instruments that can receive kinds of light

that never reach the surface of Earth. The latest is the Hubble Space Telescope, which promises to revolutionize our knowledge of the cosmos.

Giotto, the European comet mission (1986) ESA

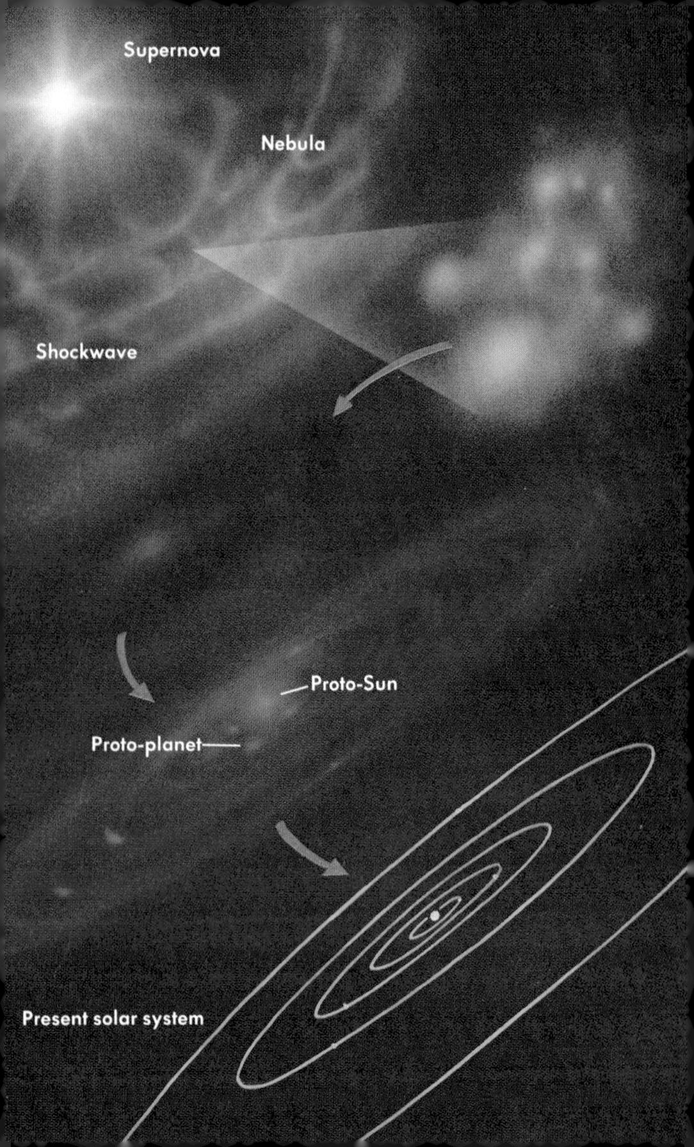

Supernova

Nebula

Shockwave

Proto-Sun

Proto-planet

Present solar system

FORMATION OF THE SUN AND PLANETS

Our universe began about 15 billion years ago with a tremendous explosion astronomers call the "Big Bang." The lighter elements, mostly hydrogen and helium, were formed in this explosion and spread throughout the universe. Later, most of these gases collected together into large bodies we call the galaxies. Our Milky Way galaxy is one of these, a giant pinwheel of stars, gas, and dust over 100,000 light-years across. Our Sun is one of several hundred billion stars in the Milky Way.

The first stars to form in our Milky Way galaxy, perhaps 10 billion years ago, were almost pure hydrogen and helium, and could not have formed hard, rocky planets because there were no heavier elements. Some of the more massive original stars lived out their lives and exploded as *supernovas*. In the extreme heat of these explosions, heavier elements such as carbon, oxygen, silicon, and iron were formed and spewed out to enrich the remaining gas.

ABOUT 4.5 BILLION YEARS AGO one such supernova caused a cloud of interstellar gas and dust to condense, and the cloud continued to collapse under the force of its own gravity. As it shrank, it grew hotter and began to spin faster. The inner portion shrank faster, and got even hotter, spinning off and leaving behind smaller orbiting clouds of gas. These smaller clouds condensed to form large lumps of gas with some heavy material at their centers. The central blob of gas collapsed still more, until it was hot enough to cause nuclear fusion, the combining of atoms of hydrogen into helium, which releases energy.

Formation of the solar system

THE BIRTH OF THE SUN, when fusion started, created a blast of radiation that blew the rest of the surrounding gas away from the center. Smaller bodies near the Sun were stripped of their gases, while the more distant ones remained massive enough to keep theirs. Thus the inner planets are small, hard, and have little or no atmosphere; the outer planets are large and mostly primordial gas. The gravitational pulls of these large proto-planets attracted more material, and moreover they were hotter. Thus each of the large outer planets formed extensive satellite systems—with inner satellites that are rocky and dense, and outer moons that are icy and less dense. In the region just beyond proto-Mars, the gravity of proto-Jupiter kept any large bodies from forming, leaving behind the asteroid belt. Comets are probably the remains of material that didn't get caught by the proto-planets.

THE PLANETS and many of their satellites later became hot enough so that they partially or totally melted, letting denser elements sink to form a core and lighter elements float up to form a crust. This heat came from several sources. One source was the heat created by their contraction from a giant sphere of gas to a smaller object. Another was heat from radioactive decay of elements such as uranium and thorium. And a third source was the energy from impacts of smaller bodies falling onto them, converting their energy of motion into heat.

Over eons, some satellites have escaped their planets, others have collided (some re-formed afterwards), and some minor objects·have been captured by the major planets to become moons. Thus the solar system was born and evolved to what we see now.

Formation of the planets

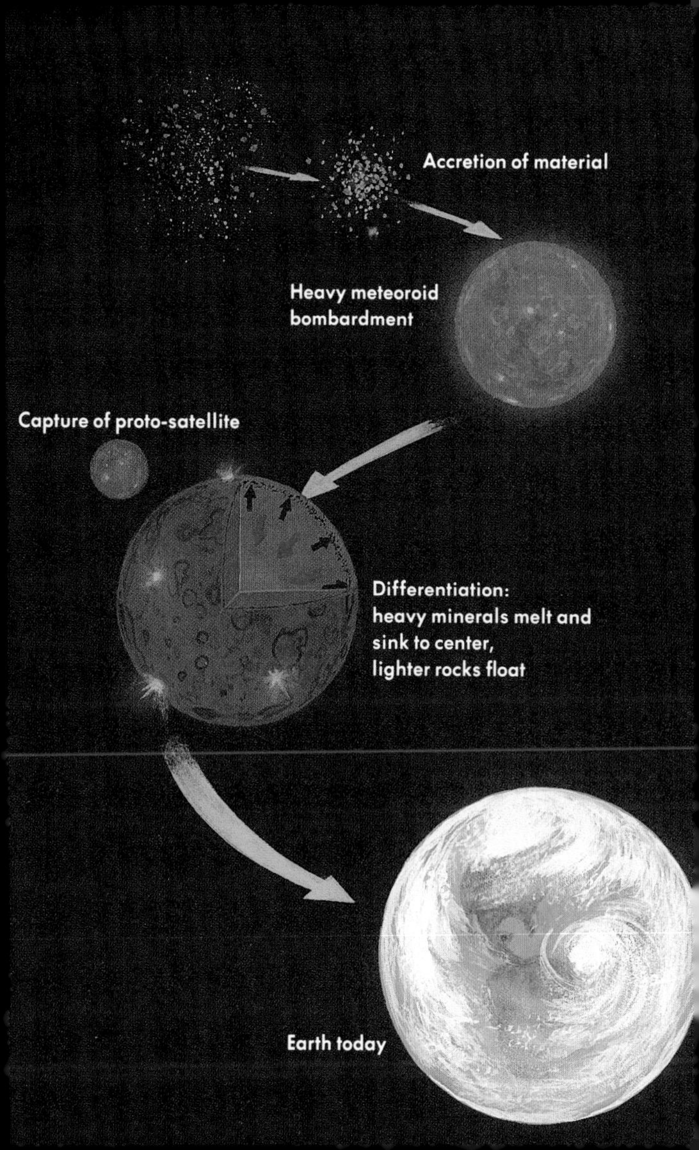

Accretion of material

Heavy meteoroid
bombardment

Capture of proto-satellite

Differentiation:
heavy minerals melt and
sink to center,
lighter rocks float

Earth today

PLANETARY ORBITS

Astronomers use the *astronomical unit*, *a.u.*, the average distance of approximately 93 million miles between the Earth and the Sun, as a unit of measure. It is almost impossible to show both the sizes of the planets and their orbits to the same scale. If the Sun really were the size shown in the illustration on the facing page, the Earth to the same scale would be 4 inches away, and Pluto would be 14 feet away.

Most planet orbits have an *inclination* within a few degrees of Earth's orbit, called the ecliptic. All the planets orbit in the same direction: Counterclockwise as seen from far above the North Pole of Earth. Planetary orbits are almost circles, except for Pluto. A number describing the amount an orbit differs from a perfect circle is called *eccentricity*.

The planets, with the

Planets compared in size

Pluto

Neptune

Uranus

Saturn

Jupiter

• Mars

● Earth

◐ Venus

• Mercury

—Sun

28

exception of Venus and Uranus, all spin on their axes in the same counterclockwise direction. Most of the satellites of the planets revolve in the same way.

PLANETARY SIZES

The planets range in size from tiny Pluto, only 1,429 miles in diameter and with a mass 0.003 times that of Earth's, to giant Jupiter, 88,732 miles across and with a mass 318 times Earth's. The Sun, with a mass of 2 billion billion billion metric tons, contains more than 99.99 percent of all the material in the solar system, and is 864,949 miles in diameter.

Minor bodies of the solar system range from the largest satellite, Jupiter's Ganymede, which is about 3,268 miles across, through the largest asteroid, Ceres, some 600 miles in size, down to microscopic dust particles.

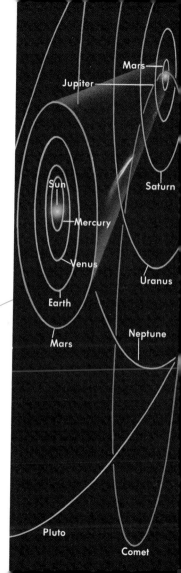

Planet orbits compared

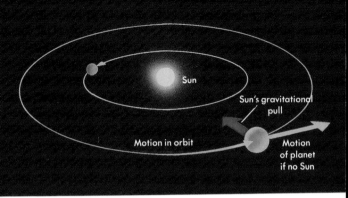

Gravity keeps planets in orbit

PLANETARY MOTION

All objects in the solar system are in motion. The planets revolve around the Sun, and the satellites revolve around the planets. At the same time, the entire solar system is moving through space, approximately toward the stars of the constellation Lyra, at a speed of 44,000 miles per hour. (But since those stars are also moving, we will never "get there.")

A PLANET'S YEAR, or *period of revolution* (also called *orbital period*), is the time it takes to orbit the Sun. The inner planets move faster, the outer ones slower. This is caused by a balancing of the force of gravity between the Sun and each planet and the planet's speed, which would tend to carry the planet away in a straight line.

A PLANET'S DAY, or *rotation period,* is the time it takes to rotate once on its own axis. While it was possible to describe the motion of planets accurately (see next page) as early as 1610, it took until the end of that century for

Sir Isaac Newton to explain why they move as they do. To accomplish that he had to discover the Law of Gravity, the Laws of Motion, and the mathematical technique called calculus.

THE LAW OF GRAVITY states that the force pulling two bodies together is proportional to the product of their masses and inversely proportional to the square of their distance apart. Mathematically, this is expressed by
$$F = G\,(M_1 \times M_2)/d^2.$$

KEPLER'S LAWS

In the early 1600s, Johannes Kepler used Tycho's observations to derive the rules by which planets (and all other bodies) move. These are summarized as *Kepler's Laws*:

I. The orbit of each planet is an ellipse and lies in a plane, with the Sun at the focus of the ellipse. (A circle is a special case of an ellipse with no eccentricity.)

II. A planet moves faster when closer to the Sun, and slower when farther away.

III. There is a fixed mathematical relationship between the size of the planet's orbit, A, expressed in a.u.'s, and the period of time it takes to go around the Sun once, T, measured in years. The relationship is $T^2 = A^3$.

Kepler's Laws

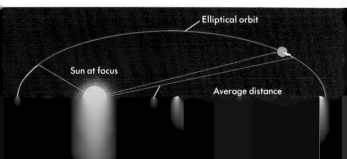

Elliptical orbit

Sun at focus

Average distance

PLANET TYPES

Astronomers often classify the planets (except Pluto, which is an anomaly) into two categories, depending on their size, structure, and composition.

TERRESTRIAL-TYPE PLANETS are those somewhat like Earth. They are Earth itself, Mercury, Venus, and Mars. Such planets are basically hard, small, rocky worlds with little or no atmosphere (even the thick atmosphere of Venus, see p. 52, is thin in comparison to the atmosphere of Jupiter). Mars' density is 3.9; the others are about 5.5.

Earth is the largest of these planets, and has the only nitrogen-oxygen atmosphere of any planet. Mercury has very little atmosphere, while the atmospheric pressure on Mars is only a few percent that of Earth's.

Before the Sun began to shine, today's terrestrial-type planets had rocky cores surrounded by hydrogen-helium atmospheres. They are all close to the Sun, so eruptions of the new-formed Sun blew away these gasses, leaving them bare rocky worlds. Over the eons, Venus, Earth, and Mars acquired their present atmospheres from gasses emitted by their rocks and from volcanic emissions. Mercury is too close to the Sun, and hence too hot, to retain much of an atmosphere.

Terrestrial-type planets also rotate comparatively slowly on their axes. Earth and Mars take about 24 hours to rotate once, Mercury takes two months, and Venus takes eight Earth months for one of its "days."

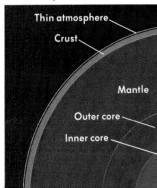

Terrestrial planet cross section

Thin atmosphere

Crust

Mantle

Outer core

Inner core

GAS-GIANT PLANETS, also called Jovian planets (after Jupiter), are large, low-density (between 0.6 and 1.6), mostly gaseous bodies. They are Jupiter, Saturn, Uranus, and Neptune. While these planets may have rocky cores (even larger than some terrestrial planets), the bulk of their material is gaseous, largely hydrogen and helium with smaller amounts of methane and ammonia. Their atmospheres, the cloud tops of which are the only thing we can observe of them, are thousands of miles thick. At the top they are very cold, hundreds of degrees below zero Fahrenheit.

Gas-giants all rotate on their axes more quickly than terrestrial planets. Jupiter and Saturn take only about ten hours for one day.

Sometimes Uranus and Neptune are put into a class of their own, intermediate between terrestrial and gas-giant types, but closer to the latter.

PLUTO does not fit either class. It is small, probably rocky and icy, has little atmosphere, and spins slowly. It is more like one of the satellites, and indeed may be an escaped satellite of Neptune.

SATELLITES of the planets, which are loosely called "moons," may be classified as either rocky or as icy, with some of them being both. A few of the satellites, most notably Titan (pp. 128-129), which is one of Saturn's 18 satellites, have atmospheres.

Jovian planet cross section

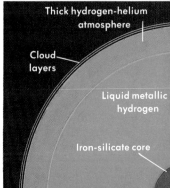

Thick hydrogen-helium atmosphere

Cloud layers

Liquid metallic hydrogen

Iron-silicate core

THE SUN

The Sun is a star and the dominant body in the solar system. It is the body that gives the solar system its name (from Sol). It is the major source of energy for all the planets, and it is by its reflected light that we are able to see them. The ancients thought the Sun a god, and gave it various names: Apollo, Ra, Sol, Helios, Shamash, Savitar. The Sun was the most important thing in the sky for the ancients, who quickly and correctly connected it with the weather, crops, and indeed their very existence. It was their time-keeper and season marker. The fundamental direction for our ancestors was the east, where the Sun comes up. Our language today recalls that when we say we "orient" ourselves.

Even as our knowledge grew, it was difficult to determine just what the Sun was, how far away it was, its size, and the origin of the Sun's energy. Not until the early 20th century were atomic fusion processes discovered that could explain the energy of the Sun and other stars.

The Sun is about 4.5 billion years old, midway through its life. In a few billion more years it will expand almost to the size of Venus' orbit and cool, becoming what astronomers call a red giant star. In the process it will end all life on Earth.

White-light image of the Sun
NASA

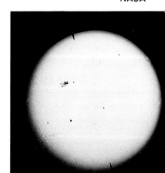

After more hundreds of millions of years, the Sun will shrink, eventually, to a small, extremely dense star the size of Earth, a white dwarf star. Then it will slowly cool over billions of years, fading to invisibility.

Prominences seen in X-ray light   NASA

The Sun undergoes very slight dimming and brightening. Even a 0.1 percent change has a noticeable effect on Earth's climate. A few percent change could have serious consequences for Earth. Fortunately, the Sun seems to be fairly stable.

When we observe the Sun in different colors of light (some of which we can't see with our eyes, like ultraviolet and X rays), we are looking at different outer layers of the Sun.

THE SUN is a ball of gas 864,000 miles in diameter. It is about 78 percent hydrogen, 20 percent helium, and 2 percent heavier elements, the most abundant of which are oxygen, carbon, nitrogen, and neon. The overall density of the Sun is only 1.4 times that of water. The Sun rotates on its axis about once a month. Its strong magnetic and electrical fields control the appearances of many of the features we observe.

The Sun in red light   NASA

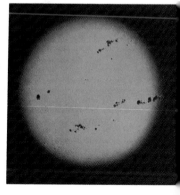

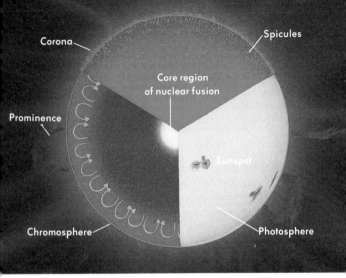

Structure of the Sun

THE PHOTOSPHERE, the "surface" we see, is a layer about 300 miles thick, with a pressure only about one-tenth that of Earth's atmosphere at the surface and a density less than a millionth of our atmosphere. The temperature of the gases in the photosphere is about 10,000 degrees Fahrenheit, which is why the Sun appears yellow. Cooler areas, which appear dark, are called *sunspots.*

INSIDE the Sun the density, pressure, and temperature increase. The outer 15 percent of the Sun is in a state of turbulent motion, like boiling water, carrying the internal energy up to the surface.

THE CORE of the Sun, only a few percent of the volume of the Sun, has temperatures of 30,000,000 degrees. The core is over eight times denser than gold, but the material is still gaseous because of the very high temperature. Here, hydrogen atoms ("H" in diagram) are combined by nuclear fusion to make helium atoms ("He") plus energy. To produce the energy we see as sunlight, more than 11 billion pounds of hydrogen turn into energy each second.

THE CHROMOSPHERE is the layer lying over the photosphere. It is about 1,000 miles thick. As we go upward, the density falls rapidly, but the temperature increases. Small jets of gas a few thousand miles high, called spicules, stick above the chromosphere.

THE CORONA, the Sun's hot, thin outer atmosphere, reaches at least a hundred million miles out into the solar system.

PROMINENCES are huge clouds of glowing gas above the photosphere. Some form like clouds and fall down to the surface. Others spring up from near the surface and may travel millions of miles into space.

THE SOLAR WIND is a high-speed flow of atomic particles out of the Sun. As they pass the Earth they are moving about a million miles per hour. Some may hit the Earth's atmosphere and cause the auroras. The solar wind can be detected far out in the solar system.

Nuclear fusion reaction

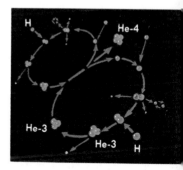

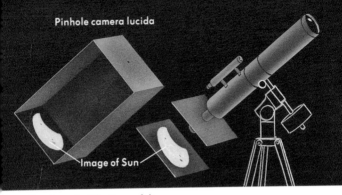

Pinhole camera lucida

Image of Sun

How to project the Sun safely

TO VIEW THE SUN SAFELY, project an image of it with a pinhole, a long focal length lens, or a telescope, onto a piece of white paper. You may use an eyepiece at the end of the telescope to make a bigger image, but do not use an eyepiece with lenses that are cemented together; to do so can ruin the eyepiece. Because all binoculars have cemented eyepieces, do not use them either. Remember: Do not look through the lens, telescope, or binoculars. To do so can result in permanent eye damage, even blindness. Look only at the projected image.

A SOLAR FILTER can be made by sandwiching two or more layers of fully exposed and developed black-and-white film

Making a sun filter

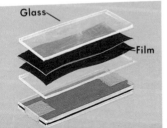

Glass

Film

negative (not color film) between two pieces of glass, and taping the edges together. Never stare at the Sun for long even through such a filter. Photographic filters are not safe for this.

SUNSPOTS, the main observable features on the

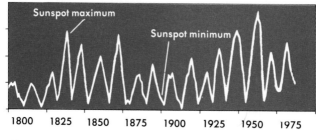

Sunspot maximum

Sunspot minimum

1800 1825 1850 1875 1900 1925 1950 1975

The sunspot cycle

Sun, are cooler areas in the photosphere. The darker, inner part of a sunspot is called the *umbra*; the outer, lighter part the *penumbra*.

The number of sunspots varies, reaching a maximum on average every 11 years. This is called the *sunspot cycle*. The next predicted sunspot maximum is in the early 1990s, and the next minimum is expected near the turn of the century.

Each day, as the Sun rotates on its axis, you may observe new sunspots on the western limb (edge) of the Sun. Day by day they are carried off eastward, to disappear off the eastern limb. Sunspots may last a few days or even weeks if they are large.

Sunspots show solar rotation

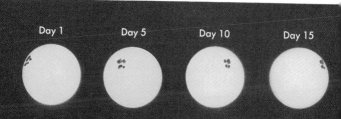

Day 1   Day 5   Day 10   Day 15

Total eclipse of the Sun   NASA

SOLAR ECLIPSES

A total eclipse of the sun is among the most dramatic and beautiful sights in nature. It occurs when the Moon moves directly in front of the Sun, hiding the light from the Sun's surface (the photosphere) and allowing us to view the faint, ghostly corona, or outer atmosphere, of the Sun. If a total solar eclipse occurs anywhere near you, try to go and see it. It will be one of the most awesome experiences of your life.

The Sun is 400 times larger than the Moon, but it is also about 400 times farther away from Earth. Thus these two objects appear to occupy the same angle in the sky, called *angular size*. The Earth is the only planet in the solar system whose satellite is exactly the same angular size as the Sun as seen from the planet's surface.

There are at least two solar eclipses each year, and some years as many as five. A minimum of zero and a maximum of three are *total eclipses*. A total eclipse can be seen only from inside a narrow band along the Earth, 160 miles wide at most, which is the track of the Moon's inner shadow, the *umbra*. It may be thousands of miles long. If you were to stay in one location, you would see, on average, a total eclipse about once every 350 years. Most people have to travel to see a total eclipse.

Outside the *path of totality*, observers see a *partial eclipse*, in which the Sun is not completely covered. Even a 99 percent partial eclipse is not anywhere near as spectacular as a total one.

Sometimes the umbra does not touch the Earth. Then there is only a partial eclipse. If the Moon's shadow is too short to reach the Earth, the Moon may appear too small to cover the Sun completely, and at mid-eclipse there will be a ring of light, or *annulus*, around the Moon. This is called an *annular eclipse*.

Solar eclipses occur only at the times of new moon, when the Moon is between the Sun and Earth. An eclipse does not occur each

Solar eclipse geometry

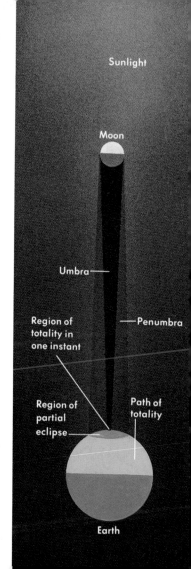

Sunlight

Moon

Umbra —

Region of totality in one instant

— Penumbra

Region of partial eclipse

Path of totality

Earth

TOTAL SOLAR ECLIPSES

Date		Duration of Totality	Region of Visibility
1990	Jul 22	2ᵐ33ˢ	Finland, Siberia, North Pacific
1991	Jul 11	6 54	Pacific, Hawaii, Mexico, Central and South America
1992	Jun 30	5 20	Atlantic Ocean
1994	Nov 3	4 24	South America, South Atlantic
1995	Oct 24	2 10	Asia, Borneo, Pacific
1997	Mar 9	2 50	Siberia
1998	Feb 26	4 08	Pacific, northern S. America, Atlantic
1999	Aug 11	2 23	Atlantic, Europe, SE and southern Asia
2001	Jun 21	4 56	Atlantic, southern Africa, Madagascar

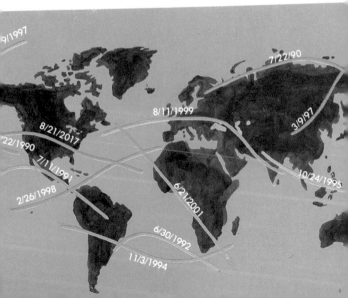

month because the orbit of the Moon is tilted compared to the path the Sun seems to take around the Earth. Eclipses thus occur in "eclipse seasons" that are about 5½ months apart.

A solar eclipse begins when the Moon just touches the Sun, called *first contact*. For about an hour of partial eclipse the Moon moves to cover more and more of the Sun. You can observe it with a filter or by projection (see p. 38). The beginning of totality, called *second contact*, is the instant the last remaining sunlight may stream through a valley on the Moon's edge, producing the beautiful "diamond ring" effect. During totality all of the Sun is covered, and it is safe to look at the total eclipse with your unaided eye. You will see the filmy corona surrounding the dark Moon, and "Bailey's Beads" of prominences around the rim. The length of totality can be from just a few seconds to slightly over 7 minutes. When totality is over, at *third contact*, there is another diamond ring, the Sun appears again, and you should again use a filter for protection. For another hour of partial eclipse the Moon slowly uncovers the Sun, and as it leaves the Sun, at *fourth contact*, the eclipse is over.

Eclipses are not only awesome, but useful as well. We have records of solar eclipses going back more than 2,500 years. They are so spectacular that many civilizations kept accurate records of them. Since eclipses can be accurately predicted as to time and areas of visibility, they provide reliable markers of time throughout history. Thus they are of use to historians in dating other events, such as the reigns of kings and dates of battles. Astronomers use them to study details of the motion of the Moon and Earth.

Paths of upcoming total solar eclipses

Surface of Mercury

MERCURY

Known to the Greeks as Hermes, Mercury of Roman mythology was the messenger of the gods and the god of commerce and thievery. It was said that his father was the god Jupiter and his mother Maia, daughter of the Titan Atlas. The Teutonic version of this god, Woden, gives his name to Wednesday. The symbol for the planet is a stylized caduceus, the staff entwined with serpents carried by Mercury. This planet has been known since prehistoric times, at least since the third millenium B.C. The planet was given this name because it moves the swiftest of all.

Mercury moves in an orbit with an average distance from the Sun of 36 million miles, or 0.39 a.u. Its orbit is very eccentric, exceeded only by that of Pluto, and it is inclined 7 degrees to Earth's orbit. It takes only 88 Earth days to revolve around the Sun once—the length of Mercury's "year." It never gets closer than 50 million miles from

Earth. Because it is always within 27 degrees of the Sun, it is hard to see.

Mercury rotates on its axis very slowly, taking almost 59 Earth days to spin once—the length of Mercury's "day." Mercury does not keep one face always toward the Sun, as was once thought. The strong gravity of the Sun has caused Mercury's rotation and revolution to be locked together in a fixed ratio: The planet rotates three times on its axis for every two revolutions about the Sun. Because of this, and the irregular motion caused by its eccentric orbit, days and nights on Mercury must be very strange. As seen from some locations on Mercury, the Sun would rise, stop, set where it rose, then rise again and continue across the sky. Then it would set, rise, and set again.

Several times during the past few centuries, astronomers claimed to have observed a planet closer to the Sun than Mercury. One astronomer named it Vulcan, and it was thought to explain why Mercury's motion is not exactly in accord with Newton's and Kepler's laws. The existence of such a planet has never been confirmed, and later astrophysicists showed that the Theory of Relativity explains the small discrepancies in Mercury's orbit.

Mercury in orbit

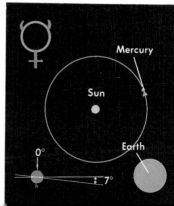

MERCURY FACTS

Distance to Sun 35.98 million miles or 0.387 a.u.
Length of year 87.97 Earth days
Orbit eccentricity 0.206
Orbit inclination 7.0 degrees
Diameter 3,031 miles or 0.382 × Earth's
Mass 0.055 × Earth's
Density 5.43
Gravity 0.38 × Earth's
Length of day 58.646 Earth days
Tilt of axis 0.0 degrees

Mercury as seen by Mariner 10   NASA

THE SURFACE OF MERCURY is much like that of Earth's Moon. There are three major kinds of terrain: smooth plains, cratered plains between large craters, and heavily cratered land. These features are not evenly distributed over Mercury's surface. Several very large features dominate the planet. Most conspicious is the Caloris Basin, which appears to be the remains of a huge impact. Concentric rings seem to be frozen ripples in the surface. Other smaller features include cracks and cliffs that are the result of the planet cooling, shrinking, and cracking.

Because the planet rotates slowly, and there is no atmosphere to carry heat around the planet, the Sun-facing side is very hot, reaching several hundred degrees Fahrenheit. The coldest measured temperature, on the side facing away from the Sun, is about 300 degrees below zero.

Mercury is only a little over 3,000 miles in diameter, about 38 percent the diameter of Earth. Mercury's density is high, about 5.4 times that of water, about equal to Earth's. This implies that Mercury has a large core of iron, which may occupy the inner three-quarters of the planet. Because of its mass, Mercury has a surface gravity such that a 100-pound person on Earth would weigh 38 pounds on Mercury. It has no known satellites.

Astronomers believe that there were five stages in the evolution of Mercury. 1) The planet's materials came together from the solar nebula, and partial melting allowed the heavier elements, notably iron, to sink to the center. 2) The planet was bombarded with meteoroids, which eroded some of the early features. 3) The largest feature on the planet, the Caloris Basin, was made by a huge meteoroid impact. 4) Lava flooded the Caloris Basin and other areas, smoothing the surface. 5) Occasional meteoroids hit the smooth plains, producing light cratering.

Mercury has a slight magnetic field, but it is not known if this is the remains of earlier magnetism, or if it is caused by dynamolike motions in the molten core. The magnetic field is not strong enough to repel the solar wind during solar storms, so the surface is constantly bombarded by atomic particles.

Close-up of Mercury   NASA

No Mercurian surface features are visible from Earth. Our only close-up photographs of Mercury's surface came from the Mariner 10 spacecraft which flew by the planet three times in 1974-75. Taking more than 10,000 photographs, Mariner 10 mapped about 57 percent of Mercury's surface.

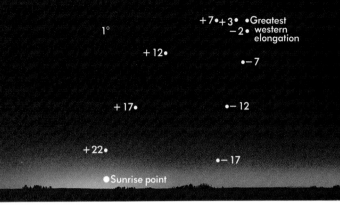

Finding chart for Mercury in the morning sky (above) and in the evening sky (opposite page)

OBSERVING MERCURY is difficult because it is never more than 27 degrees from the Sun, and thus always seen within about an hour of sunrise or sunset, while the sky is still light. The best times to see Mercury are the few days just before and after greatest elongations, denoted in the table by "E" for greatest eastern elongation (Mercury as an evening "star"), and by "W" for greatest western elongation (Mercury as a morning "star").

The illustrations show typical positions of Mercury several days before (minus) and after (plus) greatest elongations, relative to the point along the horizon where the Sun rises and sets.

TRANSITS OF MERCURY—when the planet is seen to cross the face of the Sun—occur about 13 times a century, when Mercury is at inferior conjunction near the time its orbit crosses the ecliptic. Transits in this decade occur on November 6, 1993, and November 15, 1999.

Greatest eastern elongation
• −2
• −8
• −12
+3•
+5•
• −14
+7•
• −16
+9•
• −18
• −20
+11•
+13•
• −22
+15•
• −24
+17• ● Sunset point

BEST TIMES FOR VIEWING MERCURY

1990	Feb	1	W	Jun	17	E	Oct	3	W
	Apr	13	E	Aug	4	W	Dec	15	E
	May	31	W	Oct	14	E	1997	Jan	24	W
	Aug	11	E	Nov	22	W		Apr	6	E
	Sep	24	W	1994	Feb	4	E	May	22	W
	Dec	6	E		Mar	19	W	Aug	4	E
1991	Jan	14	W	May	30	E	Sep	16	W
	Mar	27	E	Jul	17	W	Nov	28	E
	May	12	W	Sep	26	E	1998	Jan	6	W
	Jul	25	E	Nov	6	W		Mar	20	E
	Sep	7	W	1995	Jan	19	E	May	4	W
	Nov	19	E		Mar	1	W	Jul	17	E
	Dec	27	W	May	12	E	Aug	31	W
1992	Mar	9	E	Jun	29	W	Nov	11	E
	Apr	23	W	Sep	9	E	Dec	20	W
	Jul	6	E	Oct	20	W	1999	Mar	3	E
	Aug	21	W	1996	Jan	2	E		Apr	16	W
	Oct	31	E		Feb	11	W	Jun	28	E
	Dec	9	W	Apr	23	E	Aug	14	W
1993	Feb	21	E	Jun	10	W	Oct	24	E
	Apr	5	W	Aug	21	E	Dec	3	W

Surface of Venus

VENUS

The goddess of love and beauty gave her Latin name to this the brightest of the planets. To the Greeks this was Aphrodite, and before them it was Ishtar to the Babylonians. For many years, Venus was thought to be two planets: Eosphorus, the morning star, and Hesperus, the evening star. Venus, which becomes closer and brighter than any major celestial body other than the Sun and Moon, has been known since prehistoric times. The symbol for Venus is supposed to be a hand mirror.

The orbit of Venus is almost circular, 67 million miles from the Sun, or 0.72 a.u., bringing Venus within 26 million miles of Earth at times of inferior conjunction. It takes 225 Earth days to revolve around the Sun once (Venus' "year") at a speed in orbit of 22 miles per second. Venus' orbit is inclined about 3 degrees to the ecliptic. It is by far the slowest-rotating planet, taking 243 Earth days to turn once (Venus' "day").

Venus comes closer to Earth than any other planet, about 25 million miles. It takes 584 days to complete a synodic period, during which all its phases, seen from Earth, are repeated. It has no natural satellites. Venus has been explored from Earth by telescopes, which see only the tops of the clouds, and by radar, which can penetrate its thick clouds. U.S. and Soviet spacecraft have investigated Venus, and a few have landed on its surface. Others still orbit the planet, mapping its surface with radar.

In the past, science fiction writers have imagined its surface to be everything from a primeval swamp where dinosaurs roam, to totally water-covered land, to a desert. The last is closest to the truth. Because the planet is just slightly smaller than Earth, it was once called our sister planet. We now know that size is about the only similarity it has to Earth.

VENUS FACTS

Distance to Sun 67.23 million miles or 0.723 a.u.
Length of year 224.70 Earth days
Orbit eccentricity 0.007
Orbit inclination 3.4 degrees
Diameter 7,521 miles or 0.949 × Earth's
Mass 0.815 × Earth's
Density 5.24
Gravity 0.91 × Earth's
Length of day 243.017 Earth days
Tilt of axis 177.3 degrees

Venus in orbit

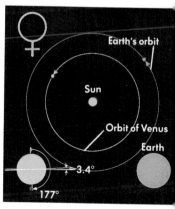

Venus seen by spacecraft in ultraviolet light   NASA

VENUS is much like the ancients imagined hell to be. On the rocky surface the temperature is 900 degrees Fahrenheit. There is little variation across the planet or from season to season. At that temperature there can be no liquid water.

THE ATMOSPHERE is the cause of the high temperature. It is 96 percent carbon dioxide, about 3 percent nitrogen, and 1 percent gases such as argon, water vapor, sulfur dioxide, and carbon monoxide. At the surface the atmosphere has 90 times the pressure of Earth's atmosphere. This dense layer of gases holds in the heat Venus receives from the Sun. The thick clouds are made mostly of hydrochloric and sulfuric acid droplets.

Wind speeds at the surface are only a few miles per hour, but at high altitudes reach several hundred miles per hour, blowing from east to west. Despite the fact that Venus has no magnetic field, it seems to have auroras high in its atmosphere; their origin is not understood.

THE AXIS OF VENUS is tilted so that its north pole is oriented almost the way the south poles of most of the other planets are. This means that the Sun rises in the west and sets in the east—once every 243 days. However, the Sun would be distorted and hard to see from the surface of Venus because of the thick atmosphere.

Venus is about 7,500 miles across, and has a density similar to the other terrestrial planets, 5.2 times that of water. It is 81 percent as massive as Earth. A 100-pound person on Earth would weigh 91 pounds on Venus.

The geology of Venus is unlike Earth's. There are craters, some from the impact of meteoroids and some from volcanoes, some of which may still be active. Both impacts and vulcanism have altered Venus' surface during the last 4.5 billion years.

MAJOR SURFACE FEATURES are named after famous real or fictional women. About a fifth of the surface is low plains, about 70 percent rolling uplands, and about 10 percent highlands. One of the largest features, Ishtar Terra, is larger than the United States and more than two miles high. The largest highland, called Aphrodite Terra, is half the size of Africa. The highest mountain, Maxwell Montes, is seven miles high, higher than Mt. Everest on Earth. Many long and wide canyons that are miles deep scar the surface. Most of Venus has been mapped by radar.

The very few photographs of the Venusian surface, taken by Soviet spacecraft, show plains of rocks stretching in all directions.

Given the surface temperature and pressure on Venus, there can be no liquid water and thus no living organisms.

1°

+60 +50 +40 +30
+70
+80
+90
+100
+110
+120
+130
+140
+150
+160
−80°
+170

+20
+10
Greatest western elongation
−10
−20
Brightest −30
−40
−50
−60
−70
Sunrise point

Finding chart for Venus in the morning sky 4/1 to 12/7, 1993

OBSERVING VENUS is easy because it is so bright. It gets close enough to us so that through good binoculars or a telescope we can see it as a small disk, and watch it change phase. The best times to observe Venus are the couple of months before and after greatest elongations. When Venus is brightest, about a month after greatest eastern elongation, and a month before greatest western elongation, it is a slim crescent.

Venus never gets more than 47 degrees from the Sun, but that allows it to remain in the evening sky well after sunset and to rise well before sunrise. It is often the first or last celestial object seen during the night. Venus remains a morning or evening "star" for seven months.

Illustrations above show the changing position of Venus in the morning and evening sky relative to the point along the horizon where the Sun rises or sets. (Remember: this point moves along the horizon as seasons change.) The numbers refer to Venus' position days before (minus) or after (plus) the greatest elongations. These are plotted for specific dates this decade, but other elongations will be similar.

54

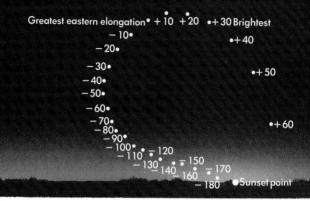

Finding chart for Venus in the evening sky 7/23/1992 to 3/30/1993

TRANSITS OF VENUS—when the planet is seen from parts of the Earth to pass across the disk of the Sun—occur very rarely, when Venus is at inferior conjunction. We are fortunate that transits will occur early in the 21st century, on June 8, 2004, visible in Europe, and on June 6, 2012, visible in Asia. The last pair of transits were in 1874 and 1882. The next will occur in 2117 and 2125.

PHENOMENA OF VENUS

Inferior Conjunction		Greatest Western Elongation		Superior Conjunction		Greatest Eastern Elongation	
1990	Jan 18	1990	Mar 30	1990	Nov 1	1991	Jun 13
1991	Aug 22	1991	Nov 2	1992	Jun 13	1993	Jan 19
1993	Apr 1	1993	Jun 10	1994	Jan 17	1994	Aug 24
1994	Nov 2	1995	Jan 13	1995	Aug 20	1996	Apr 1
1996	Jun 10	1996	Aug 20	1997	Apr 2	1997	Nov 6
1998	Jan 16	1998	Mar 27	1998	Oct 30	1999	Jun 11
1999	Aug 20	1999	Oct 30	2000	Jun 11	2001	Jan 17

Planet Earth as photographed by Apollo 16 astronauts   NASA

EARTH

Our planet is the only one to have oceans and a breathable atmosphere. And it is the only planet not named after a Greek or Roman god. The name comes from Old English and its predecessor Germanic languages.

Earth is also just the right distance from the Sun so that its average temperature is between the freezing point and the boiling point of water. Liquid water is a major condition that makes life possible.

Life on Earth began around 3 billion years ago, when primitive chemicals combined in such a way that they could take energy from their environment and reproduce themselves. Over the eons organisms grew more complex. Humans arose a few million years ago.

EARTH'S MOTIONS give us our major units of time. The *period of revolution* around the Sun is the year (also called the planet's *orbital period*). The time it takes to spin on our axis, or *period of rotation*, is the day.

Some of the things we have discovered from our close study of Earth help us understand our solar system neighbors. It is very useful to study the Earth from space, for in that way we can see things on a large scale that we recognize only with difficulty from the ground.

Earth generates a strong magnetic field that extends out into space where it interacts with the solar wind. Atomic particles travel down this field to strike the atmosphere and cause auroras. Others are trapped in donut-shaped regions around the Earth called the Van Allen belts.

EARTH FACTS

Distance to Sun 92,917,931.7 miles or 1.000 a.u.
Length of year 365.2422 days
Orbit eccentricity 0.017
Orbit inclination 0.00 degrees
Mass 1.000 × Earth's

Diameter 7,926 miles (through the equator)
Density 5.52
Gravity 1.000 × Earth's
Length of day 23ʰ 56ᵐ 04ˢ
Tilt of axis 23.4415 degrees

Earth and its magnetic field

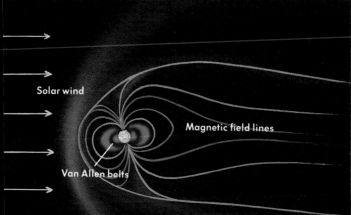

Solar wind

Magnetic field lines

Van Allen belts

EARTH'S OCEANS would be the most conspicuous feature of the planet to a visitor from space. They cover about 70 percent of the area of the globe, and if all the water were spread out, its average depth would be almost 2½ miles.

Of the water on Earth, most is seawater. If 12 gallons represented all the water on Earth, only 5 cups of that would be fresh water, and 3½ cups of that would be frozen in glaciers. Of the remainder, not quite 1½ cups is either too deep, too polluted, or exists as water vapor in the atmosphere, leaving just 6 drops of useable fresh water.

SEAWATER is about 3.5 percent salt, mostly sodium chloride, like our table salt. Small amounts of elements such as calcium, magnesium, and much smaller amounts of other elements also are in solution.

More than 80 percent of all photosynthesis, the turning of sunlight and nutrients into plant matter, takes

place in the oceans, making them the principal abode of life on Earth.

OCEAN CURRENTS, such as the Gulf Stream, carry huge amounts of the Sun's energy, which falls mostly on equatorial regions, into the temperate zones of the planet. Five major circulation patterns and 34 named major ocean currents affect climates all over the globe. There is a complex interaction with the atmosphere that produces the climate we experience on our planet.

The Earth's oceans with water removed

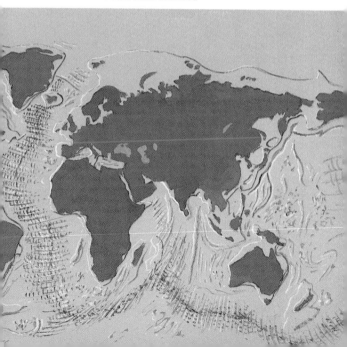

EARTH'S ATMOSPHERE is a thin layer of gases shielding the surface from space. It is composed of 78 percent nitrogen, 20 percent oxygen, and small but significant amounts of carbon dioxide, argon, and water vapor, plus traces of other gases. A column of air 1 inch square reaching from the surface to space would weigh 15 pounds, so atmospheric pressure is 15 pounds per square inch. Pressure decreases as you go higher.

The atmosphere can be divided into layers by temperature and chemical composition. The lowest layer is the troposphere, extending up about 8 miles. This is where all weather occurs. Next is the stratosphere, extending up about 30 miles. Above that, the mesosphere rises to about 50 miles, and above that is the thermosphere. The outermost layers of the atmosphere slowly thin out to become space. Above about 60 miles the air is so

Structure of Earth's atmosphere

Space

Aurora

Cosmic ray

Lowest satellit

Meteors

Ionospher

Mesosphere

Stratosphere

Troposphere

thin that a satellite can make at least one orbit without being slowed down by air drag. Many consider this the boundary between atmosphere and space.

Between 8 and 30 miles up the oxygen is in the form of ozone, having three oxygen atoms per molecule. Ozone absorbs harmful ultraviolet light and prevents it reaching Earth's surface. Some chemicals drift up from the surface and reduce the amount of ozone, allowing more dangerous ultraviolet light to reach Earth. Above 40 miles, sunlight hits atoms and knocks electrons from them, ionizing them, and creating the ionosphere. Radio waves reflected from this layer travel around the world.

There is no "top" of the atmosphere. The density of material just gets lower as you go higher. Although there is no legal definition of where the atmosphere ends and "space" begins, it is usually taken to be about 60 miles up.

AURORA BOREALIS in the northern hemisphere and *aurora australis* in the southern hemisphere are produced at the top of the atmosphere when atomic particles from the Sun strike molecules of the air and cause them to glow. These are best seen from high latitudes, but occasionally are seen closer to the equator.

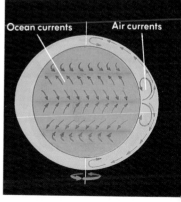

Atmospheric circulation patterns

Ocean currents   Air currents

In the lower layers, currents of air carry heat around the planet. The main heat carrier is water vapor. Differences in temperature in different parts of the globe cause most of the weather we experience.

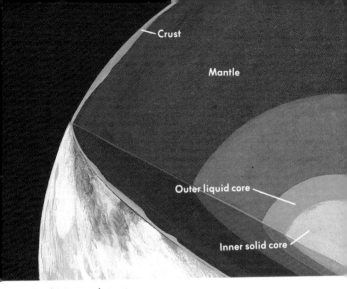

Earth's internal structure

EARTH'S INTERIOR is a place we can only explore remotely. Even our deepest mines and drills only penetrate a few miles inward toward the core, which is almost 4,000 miles below. As we go deeper, temperatures increase. The heat inside the Earth is due to the radioactive decay of such elements as uranium and thorium, plus heat left over from Earth's formation.

THE CRUST of the Earth is as thin as the skin of an apple in comparison. Its main ingredients are silicon and oxygen, which make up many materials. Continents and other "plates" of this lighter material float on the mantle, somewhat like icebergs on water. They slowly move, sometimes colliding with one another, a process called plate tectonics.

THE MANTLE lies below the crust. It is denser rock that is somewhat plastic, kept from melting by the high pressure from overlying rocks. When this rock finds a weak place in the crust, it may liquify, becoming magma. Magma that reaches the surface through volcanic activity is called lava. Rising magma along the ridges at the centers of the oceans produces more crust and pushes the plates apart. Along ocean edges, the plates collide and slip under one another, melting back into the mantle. Such regions produce earthquakes, volcanoes, and mountains.

THE EARTH'S CORE is liquid in its outer regions, composed mostly of iron and nickel, like some meteoroids. Motions in the iron core act like a dynamo to produce Earth's magnetic field.

At the very center of the Earth is the inner core, probably solid nickel-iron. The temperature at the core is around 10,000 degrees Fahrenheit.

SEISMIC WAVES, produced by earthquakes, travel differently through different kinds of rock. By analyzing the waves, geologists can deduce the internal structure of the planet. Earthquakes are caused by breakage and slippage of rocks inside the Earth along weak regions called faults. Hundreds of microquakes occur daily. Large earthquakes are rare, but can be very destructive.

Volcanic activity   USGS

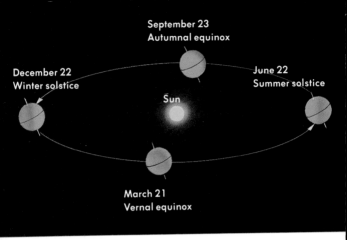

Earth orbiting the Sun

THE SEASONS

Because the Earth is tilted on its axis by 23½ degrees, different parts of the planet get different amounts of heat from the Sun.

When the northern hemisphere is tilted toward the Sun, the days there are longer and the Sun is high in the sky. Then that hemisphere gains more heat in the day than it radiates away into space at night. Hence the weather grows warm, and we have summer.

Six months later, the northern hemisphere is tilted away from the Sun, days are short, and the Sun is low in the sky. More heat is lost at night than is gained, and temperatures fall. It is winter.

It is this tilted axis, not the slight changing distance between Earth and Sun, that makes the seasons. Earth is closest to the Sun about January 3, and farthest from the Sun, about 3 percent farther away, about July 4 each year.

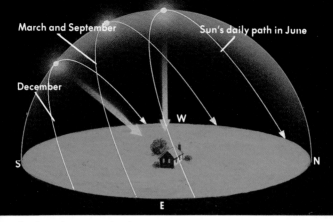

Sun's motion across the sky

Seasons in the southern hemisphere are the opposite of those in the northern hemisphere.

THE SOLSTICES are the times when the Sun is farthest north or south. The *summer solstice* (for the northern hemisphere), about June 21, is the beginning of summer. The *winter solstice*, about December 22, is the start of that season.

THE EQUINOXES are the times when the Sun crosses the equator. The northward crossing, the *vernal equinox*, is about March 21, the beginning of spring. The *autumnal equinox* is about September 23.

The hottest and coldest months do not correspond to the Sun's northern or southern extremes because it takes time to heat up and to cool off the land, the atmosphere, and the oceans.

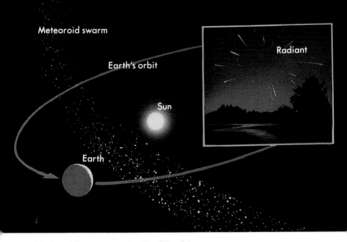

Meteoroid swarm crossing Earth's orbit

METEORS are streaks of light seen in the night sky; they are caused by small particles from space whizzing into the upper atmosphere at speeds of 25 miles per second. The friction of their motion heats the air around them to incandescence. On a typical dark night, about five to ten meteors per hour can be seen. About ten tons of such particles hit the Earth each day. Most are only the size of a grain of sand. Even bright meteors are caused by particles only the size of a small pea. Rarely, very bright meteors called *fireballs,* or *bolides* if they explode, can be seen even in daylight.

METEOROIDS are the particles while they are still in space. Most are debris from comets and are made of *stony* material. Others are a nickle-iron alloy, called *irons*. A few rare ones are a mixture of *stone and iron.*

METEORITES are meteoroids that have hit the ground. Most are small. A very few have weighed tons, and very rarely, perhaps once every few tens of thousands of years, a huge crater may be formed. Although the most common type of meteorite to fall are stony, the most common meteorites found are the iron ones, as they are more conspicuous on the ground and resist erosion.

METEOR SHOWERS occur when the Earth passes through swarms of meteoroids that are following in the orbits of comets. At such times, a large number of meteors can be seen, up to about 60 an hour, and occasionally more. Seen from Earth they seem to radiate from a small region of the sky, called the *radiant,* named after the constellation in which their radiant is located. The number of meteors seen gradually rises a few days before and diminishes for a few days after the peak date.

TABLE OF MAJOR METEOR SHOWERS

Peak Date	Name (radiant)	Number per Hour	Duration (in days)
Jan 4	Quadrantids	40	2.2
Apr 21	Lyrids	15	4
May 4	Eta Aquarids	20	6
Jul 28	Delta Aquarids	20	14
Aug 12	Perseids	50	9.2
Oct 21	Orionids	25	4
Nov 3	South Taurids	15	?
Nov 16	Leonids	15	?
Dec 13	Geminids	50	5.2
Dec 22	Ursids	15	4

Note The duration is the total number of days, centered on the peak, during which the hourly rate is at least ¼ of the peak rate.

Earth's Moon   NASA

THE MOON

The Moon is the most conspicuous celestial object after the
Sun. The Moon goddess was called Luna by the Romans,
Artemis and Selene by the Greeks. "Moon" is a word from
the old Germanic languages and the Greek "mena," from
an older word meaning "to measure." It shows the early
interest in our satellite as a measurer of time. It orbits the
Earth once a month in an orbit that is an average of
239,000 miles from us, moving at about 2,000 miles per
hour. The word "month" comes from "moneth," an early
word for the Moon. The Moon's closeness to Earth has
caused it to keep one face always toward Earth.

For thousands of years the Moon has marked the passage of time with its phases. Hebrew, Islamic, and Chinese lunar calendars are still in use today. Because the interval between full moons is about 29½ days, sometimes there are 12 full moons and some years 13 full moons in one calendar year of 365¼ days.

There is much folklore about the Moon. Legends claim you become lunatic sleeping in the light of the Moon. Others claim you should plant and harvest particular crops at certain phases of the Moon. Some claim crime increases at full moon. Most of this folklore has no truth to it.

THE HARVEST MOON is the full moon closest to the autumnal equinox, rising around sunset for several nights in a row. Its light helped extend hours for reaping the fields. The following full moon is often called the Hunter's Moon, for its light allowed hunters extra hours to gather supplies for the winter.

The Moon is the only other planetary body explored in person by mankind—during the U.S. Apollo program from 1969-72.

MOON FACTS

Diameter 2,159 miles or 0.27 × Earth's

Mass 0.0123 × Earth's

Density 3.34

Gravity 0.16 × Earth's

Length of day 27.3217 Earth days

Tilt of axis 6.7 degrees

Length of year
compared to the stars (sidereal): $27^d07^h43^m$
relative to Earth (phases): $29^d12^h44^m$

Average distance from Earth 238,820 miles

Orbit eccentricity 0.055

Orbit inclination (varies) 18.3 to 28.6 degrees

Lunar "sea" NASA   Lunar highlands CARNEGIE

THE MOON is about one-quarter the size and $\frac{1}{81}$ the mass of the Earth. It has no atmosphere and no water, and probably never had any. A 100-pound object on Earth would weigh 16 pounds on the Moon.

THE SURFACE OF THE MOON is covered by a layer of powder and rubble, called the regolith, produced by meteoroid impacts. Major surface features are the darker areas of smooth lava flows, the "seas" called maria (pronounced *mar-ee-ah*; singular mare, pronounced *mar-ay*); walled craters and plains; mountains and highland areas; and many, many craters, most produced by impact. The Moon's crust is about 40 miles thick on the side of the Moon facing us and about 60 miles thick on the far side. Lunar features are named for famous scientists and explorers.

Beneath the crust is the mantle, made of denser rock, extending down to about 500 miles below the surface. At this layer many small moonquakes occur that can be measured by instruments left on the surface by the Apollo astronauts. We still know little about the core of the Moon. Since the Moon has no magnetic field, the core probably does not act like a dynamo as does Earth's.

MOON ROCKS are much like those of the Earth, but enriched in some elements. Moon rocks have no water and have never been exposed to oxygen. Thus the chemical changes and erosion that altered all Earth rocks have not affected lunar materials.

Lunar mountains NASA

When astronauts establish settlements on the Moon early in the 21st century, they will use lunar rocks to construct their buildings. They will use the abundant solar power to decompose the lunar rocks into their elements, then reassemble those elements into water, air, and other necessities of life.

More than 800 pounds of lunar rocks are being studied in laboratories on Earth for clues to the origin and evolution of the planets.

FORMATION OF THE MOON is still a puzzle. There are three main theories: It was once a part of the Earth that split off; it was formed elsewhere in the early solar system and captured when it passed near Earth; it was formed by the accretion of smaller bodies where it is now. More studies, and probably visits, will be necessary to decide between these competing theories—or to develop a new one!

Lunar craters NASA

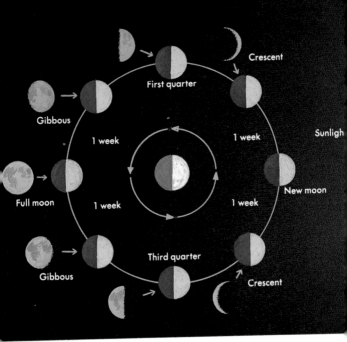

First quarter

Crescent

Gibbous

Sunligh

1 week

1 week

Full moon

1 week

1 week

New moon

Gibbous

Third quarter

Crescent

THE MOON'S ORBIT takes it around Earth once a month. During a day, the Moon moves from east to west across our skies, caused by the turning of the Earth. At the same time, its orbital motion carries it in a west to east direction against the background of stars an amount about equal to its own diameter every hour. This combination of motions means that the time of moonrise is later each day, averaging about 54 minutes. In fall the daily delay is smaller, and in spring it is greater.

PHASES OF THE MOON are the apparent changes in shape due to the changing illumination of the Moon by the Sun, as seen from Earth. Moonlight is just reflected sunlight. The far side of the Moon is not always the "dark side."

NEW MOON occurs when the Moon is approximately between Earth and Sun. Then the illuminated side is away from us, so the Moon appears dark. At such times, the Moon rises and sets with the Sun. An eclipse of the Sun can occur at new moon.

As the Moon moves east of the Sun, it appears in our sky as a crescent (the bright side always faces the Sun). Each day it rises later, the crescent is wider, and it is higher in the sky at sunset. We say the Moon is *waxing*.

FIRST-QUARTER MOON (sometimes called half moon) occurs about a week after new moon, when the Moon is a quarter of its way around its orbit. We see the bright western half of the disk. At this phase the Moon is high in the southern sky at sunset.

As days go by, the Moon becomes rounded on both sides, and is called *gibbous*. Each night it rises later, and is fuller.

FULL MOON is when the Moon is opposite the Sun in its orbit, about two weeks after new moon. It is then fully illuminated as seen from Earth, and opposite the Sun, rising about sunset and setting about sunrise. At such times there can be eclipses of the Moon. After full moon, the moon is *waning*.

LAST QUARTER (or third quarter), about a week later, is when we see the eastern half of the Moon illuminated. The Moon then rises about midnight and sets about noon. After that, it is a waning crescent, until about a week after third quarter when it is again new moon. Then the cycle repeats.

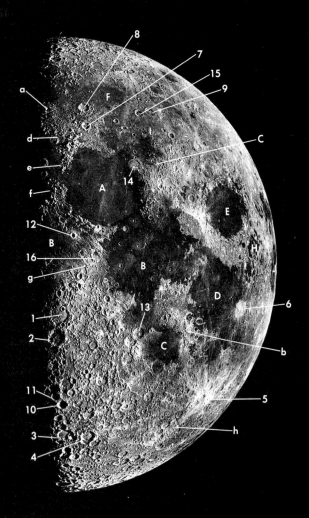

THE FIRST-QUARTER MOON

The features on the Moon are best seen not at full moon, when the sunlight is right overhead for them, but around first and last quarter, when the light comes from the side and shadows emphasize the terrain. As the Moon waxes and wanes, each night the line separating light and dark—called the *terminator*—lies across a different part of the Moon. Surface features near the terminator are especially visible.

Here are some of the major features visible on the near side of the Moon using binoculars or a small telescope.

MARIA:

A. Mare Serenitatis
B. Mare Tranquilitatis
C. Mare Nectaris
D. Mare Foecunditatis
E. Mare Crisium
F. Mare Frigoris
G. Mare Vaporum
H. Mare Undarum
I. Lacus Mortis
J. Mare Smythii

CRATERS AND WALLED PLAINS:

1. Hipparchus
2. Albategnius
3. Stöfler
4. Maurolycus
5. Petavius
6. Langrenus
7. Eudoxus
8. Aristoteles
9. Atlas
10. Aliacencis
11. Werner
12. Manilius
13. Theophilus
14. Posidonius
15. Hercules
16. Julius Caesar

OTHER FEATURES:

a. Alpine Valley
b. Pyrenees Mountains
c. Taurus Mountains
d. Alps Mountains
e. Caucasus Mountains
f. Appenines Mountains
g. Ariadaeus Rille
h. Reita Valley

THE LAST-QUARTER MOON

The third-quarter moon is less familiar to many people because it doesn't rise until late in the evening. Nevertheless, it holds some of the more interesting features visible on the Moon.

MARIA:

A. Mare Imbrium
B. Sinus Iridium
C. Mare Frigoris
D. Oceanus Procellarum
E. Mare Humorum
F. Mare Nubium
G. Sinus Roris
H. Sinus Aestuum
I. Sinus Medii
J. Mare Vaporum

CRATERS AND WALLED PLAINS:

1. Clavius
2. Maginus
3. Tycho
4. Arzachel
5. Alphonsus
6. Ptolemaeus
7. Albategnius
8. Eratosthenes
9. Archimedes
10. Autolycus
11. Epigenes
12. Timocharis
13. Copernicus
14. Aristarchus
15. Herodotus
16. Kepler
17. Alpetragius
18. Herschel
19. Pallas
20. Grimaldi

OTHER FEATURES:

a. Alpine Valley
b. Straight Range
c. Jura Mountains
d. Riphaeus Mountains
e. Straight Wall
f. Appenines Mountains
g. Alps Mountains
h. Piton Mountain
i. Pico Mountain
j. Harbinger Mountains

LICK OBSERVATORY PHOTOGRAPH

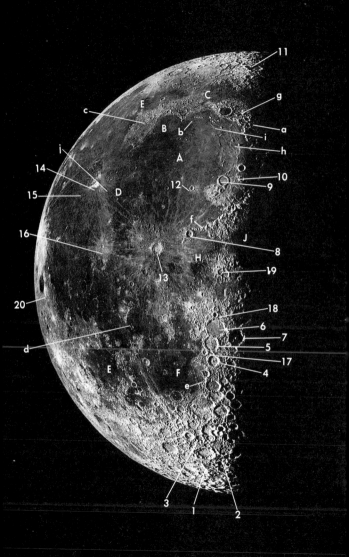

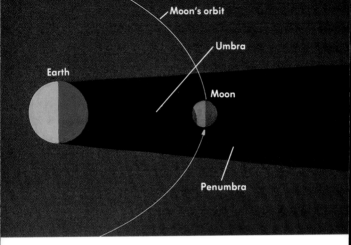

Geometry of a lunar eclipse

LUNAR ECLIPSES occur at full moon when the Moon passes into the shadow of the Earth. They do not occur each full moon because of the tilt of the Moon's orbit. Usually the Moon passes north or south of Earth's shadow.

The Earth's shadow is a cone pointing away from the Sun. You cannot see the shadow except when it falls on something, such as the Moon. At the Moon's distance from Earth, the darker inner portion of the shadow, the *umbra*, is about 5,700 miles across. The lighter, outer portion where not all of the Sun's light is cut off, the *penumbra*, is about 10,000 miles across.

If the Moon passes only through the penumbra, it is called a *penumbral eclipse*. Such eclipses are almost not noticeable, because the dimming of the Moon is slight. If the Moon passes completely into the umbra, it will be a *total eclipse*. A *partial eclipse* occurs when the Moon does not fully enter the umbra.

The Moon during eclipse   J. Wyckoff

Like solar eclipses, lunar eclipses take place during *eclipse seasons*, times of the year when the Sun appears to cross the Moon's orbit near the phases of full and new moons. Such eclipse seasons happen about every 5½ months. Thus the months of the calendar year in which eclipses occur slowly change.

A TOTAL LUNAR ECLIPSE begins when the Moon just touches the edge of the umbra, called *first contact*. From Earth, we see one edge of the Moon grow dim, and this wedge of darkness slowly spreads over the Moon. At *second contact*, the Moon is just within the umbra, and totality begins; it lasts a maximum of not quite two hours.

During *totality*, not all light is cut off. The Moon is usually a faint reddish-orange disk, illuminated by light refracted and filtered through the Earth's atmosphere. How bright it is during totality depends on how many clouds are around the sunrise-sunset line of Earth at the time.

At *third contact* the Moon begins to leave the umbra, and at *fourth contact*, the eclipse is over. The maximum possible duration of a total eclipse, from first to fourth contact, is 3 hours 40 minutes.

Lunar eclipses can also be used by historians as date markers, and have influenced historical events. It is said that Columbus used his knowledge of an upcoming lunar eclipse to frighten natives into supplying his ships.

THE FREQUENCY OF LUNAR ECLIPSES is slightly less than that of solar eclipses. They are more commonly seen, however, for they can be observed by everyone on the nighttime side of the Earth. At least two, and a maximum of five, lunar eclipses can occur each year. In the table below, "T" indicates a total eclipse, "P" a partial one.

TABLE OF LUNAR ECLIPSES

Date		Type	Areas of Visibility
1990	Feb 9	T	Arctic, Australasia, Europe, Africa
1990	Aug 6	P	SW Alaska, Pacific, Antarctica, Australasia, S and E Asia
1991	Dec 21	P	Arctic, NW South America, North America, Pacific, Australasia
1992	Jun 15	P	S and W Africa, SW Europe, Americas except the NW, Antarctica
1992	Dec 9*	T	Asia except SE, Africa, Europe, North America except W, Central and South America except S
1993	Jun 4	T	S of South America, W of North America, Pacific, Australasia, SE Asia
1993	Nov 29	T	Europe, W Africa, Arctic, Americas, N Asia
1994	May 25	P	Africa, Europe except NE, Americas except NW and extreme N
1995	Apr 15	P	W of North America, Pacific, Australasia, E and SE Asia
1996	Apr 3**	T	W Asia, Africa, Europe, South America, West Indies, E of North America
1996	Sep 27	T	W Asia, Africa, Europe, Americas except Alaska
1997	Mar 24	P	Africa except E, Europe except NE, Americas except W Alaska
1997	Sep 16	T	Australasia, Asia except NE, Africa except W, Europe
1999	Jul 28	P	W and S of South America, Central America, W of North America, Pacific, Australasia, E Asia
2000	Jan 21	T	N and NW Asia, W and N Africa, Europe, Americas
2000	Jul 16	T	Pacific, SW Alaska, Australasia, SW Asia

*May occur on Dec 10 in some locations.
**May occur on Apr 4 in some locations.

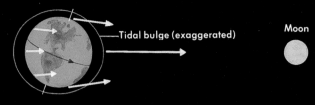

Moon

Tidal bulge (exaggerated)

Length of arrows shows strength of gravitational pull

Geometry of the tides

THE TIDES are caused when the Moon and Sun pull the water on the near side of the Earth upward away from the Earth, while the Earth itself is pulled slightly away from water on the far side. The Sun's effect on the tides is only about a third that of the Moon. High and low tides occur 12½ hours apart. Actual tides at coastal locations are greatly affected by the body of water and the shape of the bottom of the ocean near the site. Near times of new and full moon, the tides are especially high, called *spring tides*. At times of the quarter moons, the tides are lower, called *neap tides*.

Tidal variation during one month

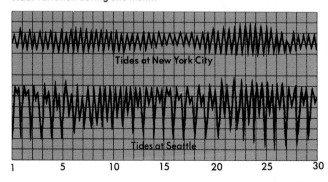

Tides at New York City

Tides at Seattle

1   5   10   15   20   25   30

Surface of Mars

MARS

The "Red Planet" has long attracted attention. Many ancient cultures associated the red color of Mars with blood, so they gave to this planet the name of the God of War. He was Ares to the Greeks. The symbol for Mars is a stylized shield and spear.

Mars has been explored by spacecraft from the U.S. and the Soviet Union. In 1976, two U.S. Viking craft soft-landed on the planet, took and analyzed soil samples, measured the atmosphere, and sent back photographs. Despite hopes, no signs of life were found.

Mars has the third most elliptical orbit. Its average distance from the Sun is 142 million miles, or 1.52 a.u. Its synodic period, or return to Sun-Earth-Mars alignment, is 2 years and 50 days. Because of its eccentric orbit, when at opposition Mars can be as much as 60 million or as little as 35 million miles from us. These closest oppositions occur about every 17 years.

Mars has a day almost the same as Earth's, 24 hours, 37 minutes. The tilt of its axis is also almost the same as Earth's, 25.2 degrees. This gives Mars day-night cycles and seasons much like Earth's. But because Mars is much farther from the Sun, the temperatures there are much lower and seasons are longer.

About a hundred years ago, astronomers thought they saw straight lines on the planet, and called them "canals." This made some people think there might be life on Mars, perhaps an ancient civilization. Modern explorations have shown the canals were a mixture of optical illusion and wishful thinking, for they do not exist.

More than any other planet, Mars has been thought a possible abode of life beyond Earth. In 1924, during a close opposition of Mars, radio stations were silenced to enable scientists to listen for radio signals from Mars. In 1938, a radio drama convinced many listeners that Earth had been invaded by Martians. Mars will certainly be the first planet to be explored in person by humans, probably within the first quarter of the 21st century.

Mars in orbit

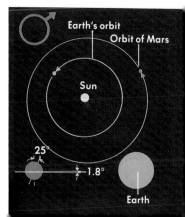

MARS FACTS

Distance to Sun 141.6 million miles or 1.524 a.u.
Length of year 686.98 Earth days
Orbit eccentricity 0.093
Orbit inclination 1.8 degrees
Diameter 4,219 miles or 0.533 × Earth's
Mass 0.1074 × Earth's
Density 3.94
Gravity 0.38 × Earth's
Length of day 24ʰ37ᵐ
Tilt of axis 25.2 degrees

Mars from low orbit   NASA

THE PLANET MARS is more like Earth than any other planet, but there are also great differences. The fourth planet is 4,217 miles in diameter, about half the diameter of Earth. Its mass is only 11 percent that of Earth's. A 100-pound weight on Earth would be 38 pounds on Mars.

THE SURFACE OF MARS is complex and changing. There are huge level areas, deserts, enormous mountains, deep canyons, and craters of all sizes. Some of these features were produced by volcanoes, some by meteoroid impacts, some probably by flowing water.

Olympus Mons   NASA

Mars has the largest volcano in the solar system. It is called Olympus Mons and is more than 16 miles high, three times taller than any volcano on Earth. Mars also has the largest valley system, a complex system of channels called Valles Marineris. There is no permanent surface water now on Mars, but perhaps there was in the past, when the climate was different. The low atmospheric pressure means free water would evaporate instantly. Some water is frozen into snow and ice as part of the northern ice cap. The southern

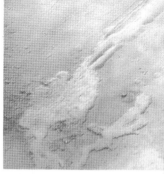

Valles Marineris, the so-called "Grand Canyon of Mars" NASA

ice cap is made mostly of dry ice, frozen carbon dioxide. More water may be trapped as ice or water deep in the soil.

Since Mars is farther from the Sun than Earth, it is much colder. At the sites of the two Viking landings, temperatures range from almost 200 degrees below zero to as high as around the freezing point of water. Some of the surface canyons and other features make scientists think that volcanic activity or meteoroid strikes may occasionally melt subsurface ice, causing huge floods. Then the water evaporates or sinks into the ground again. Despite its thin atmosphere, Mars has weather that includes cyclones and dust storms that may cover half the planet. No signs of life have been found.

Martian surface seen by Viking NASA

THE SATELLITES OF MARS are small, rocky, low-density bodies discovered in 1877, although they had been "predicted" more than a century earlier in the story of *Gulliver's Travels.* Their names are Phobos and Deimos, the Greek names for the dogs of Ares: Fear and Panic. Both are probably asteroids that came too close to Mars and were captured. Both would make ideal space stations from which to study Mars itself.

Phobos   NASA

Deimos   NASA

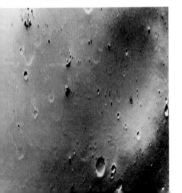

PHOBOS, the inner satellite, is 5,840 miles from Mars' center, which is the usual way astronomers measure distance of a satellite, but only 3,700 miles above the surface of Mars. It orbits Mars once every 7.66 hours, in a west-to-east direction like most of the other satellites in the solar system. This orbital period is much less than the 24.62-hour rotation period of Mars itself. Thus, if you were on Mars, you would see Phobos as a bright point of light, rising in the west and setting in the east. It would rise usually twice each Martian day. Phobos itself is about 12 by 8 miles in size, with a very dark, rocky surface pocked by craters.

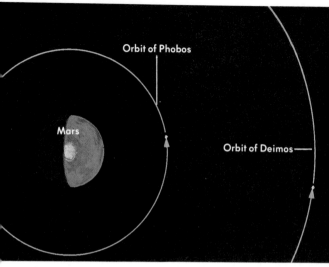

Orbits of Mars' satellites

DEIMOS orbits Mars every 1.26 days at a distance of 12,500 miles above Mars' surface. Since its period of revolution is close to the rotation period of Mars, Deimos would rise in the east and slowly move across the sky, remaining visible in the Martian sky for several Martian days before setting in the west. It is around 7 by 6 miles in size.

SATELLITES OF MARS

Name	Year Disc.	Distance from Mars (miles)	Period of Revolution (days)	Mass (x the Moon's)	Density	Size (miles)
Phobos	1877	5,840	0.319	0.00000018	1.9	12 × 8
Deimos	1877	14,602	1.263	0.000000024	2.1	7 × 6

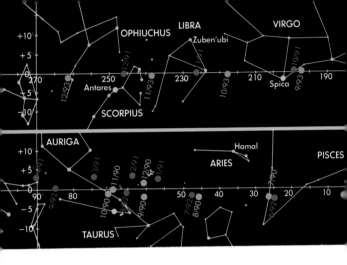

OBSERVING MARS The "Red Planet" goes around the Sun once each 687 days, or 1.88 Earth years. Mars is best seen around the times of opposition, the times it is closest to Earth. Mars' synodic period, the time between successive oppositions, is 779 days, or about 26 months. Thus it takes more than two years for Mars to circle the sky as seen from Earth.

Not all oppositions are good for viewing because the eccentricity of Mars' orbit means that at some oppositions Mars is closer to Earth than at others. These favorable oppositions, at which the distance to Earth is about 35 million miles, occur about every 17 years. The next several oppositions and Mars' distance from Earth at the time are given in the table. The finding chart above shows the position of Mars against the background of the constellations of the zodiac.

When Mars is nearest, it can be very bright and distinctly reddish in color. When distant, it is only like a medium-

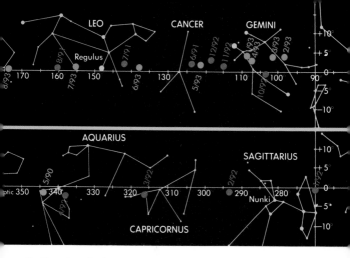

Finding chart for Mars

bright star, and may not appear red. Sometimes global dust storms in Mars' atmosphere obscure everything for weeks or months.

From Earth, Phobos and Deimos are not visible in any telescope likely to be owned by an amateur, since the satellites are very small and do not reflect much sunlight.

TABLE OF MARS' OPPOSITIONS

Date of Opposition		Constellation	Distance from Earth (miles)
1990	Nov 27	Taurus	48 million
1993	Jan 7	Gemini	58 million
1995	Feb 12	Leo	63 million
1997	Mar 17	Virgo	61 million
1999	Apr 24	Virgo	54 million
2001	Jun 13	Ophiuchus	42 million
2003	Aug 28	Aquarius	35 million
2005	Nov 7	Aries	43 million

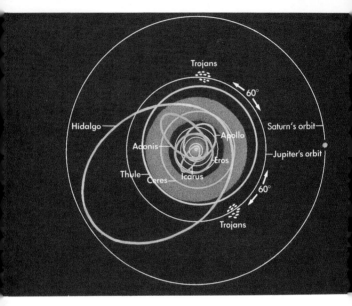

Asteroid orbits

ASTEROIDS: THE MINOR PLANETS

The asteroids (meaning "starlike"), or minor planets, are a multitude of small solid bodies, most of which are located in a huge donut-shaped region lying between Mars and Jupiter, called the *asteroid belt*, and extending somewhat above and below the plane of the ecliptic. They range in size from invisible pebbles up to the largest, Ceres, about 600 miles across. Their small size and low reflectivity account for the fact that the first one, Ceres, was not discovered until the first day of the 19th century, by Italian astronomer Giuseppe Piazzi.

A few asteroids have orbits that take them outside the asteroid belt. If an asteroid's orbit crosses within the orbit of Mars it is called an Amor-type asteroid. If it crosses within Earth's orbit, it is an Apollo-type asteroid. These are named after the first asteroid of each kind to be discovered.

As with planets, the first asteroids discovered were named for mythological figures, and they were numbered in order of discovery. When many more had been found, their discoverers were allowed to give them other names—of friends, relatives, pets, plants, or almost anything else. The full name of an asteroid consists of the number plus a given name, if any. Some examples of asteroid names are 433 Eros, 511 Davida, 8 Flora, 349 Dembowska, and 324 Bamberga. About 1,700 asteroids have names, many hundreds more only numbers. There are probably hundreds of thousands of minor planets of significant size.

In the asteroid belt

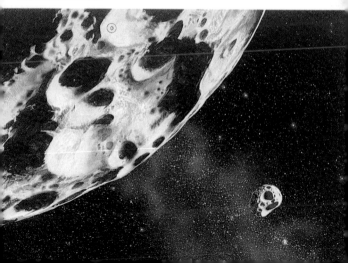

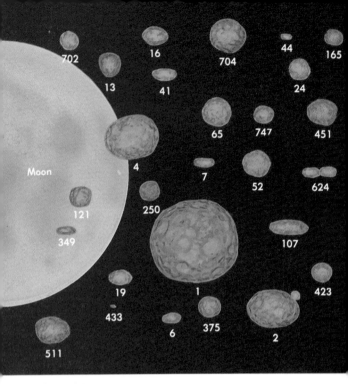

Some of the minor planets

Key to minor planets shown above:

1	Ceres	19	Fortuna	107	Camille	433	Eros
2	Pallas	24	Themis	121	Hermione	451	Patientia
4	Vesta	41	Daphne	165	Loreley	511	Davida
6	Hebe	44	Nysa	250	Bettina	624	Hektor
7	Iris	52	Europa	349	Dembowska	702	Alauda
13	Egeria	65	Cymbele	375	Ursula	704	Interamnia
16	Psyche			423	Diotima	747	Winchester

IN COMPOSITION the minor planets most closely resemble meteorites that fall to Earth. Probably many meteorites are chips knocked off asteroids by collisions among themselves. (Others come from comet debris, p. 151.)

Also like meteorites, there are several types of asteroids. Some seem to be mostly metallic (nickel-iron), while others resemble basalt (a volcanic rock found on Earth), and still others contain carbon compounds and are very dark. They may be rich in lighter elements.

Astronomers think asteroids are debris from the formation of the solar system. Because they were too close to Jupiter and its powerful gravitational field, they could not form large bodies. Also, some may be fragments of collisions between larger bodies. The few minor planets that come into the inner solar system are thought to have had their paths perturbed by collisions. There may be other belts of minor planets farther out in the solar system, but only one asteroid, Chiron, has been discovered orbiting beyond Saturn.

Minor planets rotate with periods ranging from three hours to several days. Most are single bodies and irregular in shape, and are probably pitted and scarred by collisions. Some are very elongated, and some asteroids may be double bodies, or even have smaller satellites. Most are small; only 33 asteroids are larger than 125 miles across.

KIRKWOOD GAPS are regions of the asteroid belt where no asteroids orbit, due to complex gravitational interactions of the minor planets with Mars and Jupiter. There are also groups of asteroids with similar properties, probably produced when one large body was broken up by a collision. These are called Hirayama families, after the Japanese astronomer who first identified them. Two special groups travel in Jupiter's orbit—the Trojan asteroids (p. 111).

Asteroid trail among stars

OBSERVING MINOR PLANETS is not easy because they are small and hence dim. Only one, 4 Vesta, gets bright enough to see with the unaided eye, and then just barely, under good conditions. It is possible that an asteroid could pass very close to Earth and become visible, but this would be very rare.

Asteroids can be seen in small and medium-sized telescopes, and about 40 brighter ones are sometimes visible in binoculars with 50 millimeter lenses. Even then, you must have an accurate star chart and know where to look, for they are not conspicuous. They appear just like stars. If you watch over a period of several hours, you may notice the motion of an object against the background of stars. This is the way they are discovered. The motion is easier to see in a guided telescopic time-lapse photograph where the asteroid shows up as a streak of light, while the stars are points of light.

The brightest minor planets are those that come close to Earth, called Earth-crossers, or Apollo objects, named

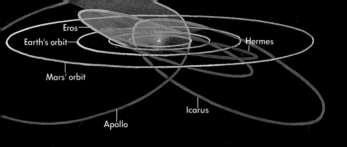

Earth-crossing asteroid orbits

after one of them. Someday these may be mined as sources of minerals and iron.

SELECTED ASTEROIDS

Name and Number	Year Disc.	Period of Revolution (years)	Average Distance from Sun
1 Ceres	1801	4.60	2.766 a.u.
2 Pallas	1802	4.61	2.768
3 Juno	1804	4.36	2.668
4 Vesta	1807	3.63	2.362
6 Hebe	1847	3.78	2.426
16 Psyche	1852	5.00	2.923
433 Eros	1898	1.76	1.458
1566 Icarus	1949	1.12	1.078
1620 Geographos	1951	1.39	1.244

View in Jupiter's atmosphere

JUPITER

The largest planet in the solar system is some 318 times the mass of Earth and 0.1 percent the mass of the Sun; it contains 71 percent of all the non-solar material in the solar system. Around it are a thin ring and an extensive system of satellites, almost a miniature planetary system.

Jupiter, known also to the Romans as Jove and to the Greeks as Zeus, was king of the gods, ruler of Olympus. The astronomical symbol for this planet is a stylized thunderbolt. The planet is among the brightest objects in the sky, and so has been known since prehistoric times. It circles the Sun once every 11.86 Earth years at an average distance of 483 million miles.

The prototype of the gas-giant planets, Jupiter is mostly gaseous and perhaps partially liquid, with at most a relatively small solid core. The "surface" we see in our telescopes is actually the top of a thick cloud layer. The planet has an average density of only 1.4 times that of water, about a quarter the density of rocky planets like Earth. Like the other gas-giants, it spins quickly on its axis: a Jovian "day," a "day" on Jupiter, lasts only 9.8 Earth hours. Jupiter is tilted 3 degrees compared to its orbit.

Jupiter's mass and size give it a tremendous gravitational field. A 100-pound object on Earth would weigh 254 pounds on Jupiter. The planet has been explored by two Pioneer and two Voyager spacecraft, which discovered many exciting details about this the largest planet.

Jupiter emits about twice as much energy as it receives from the Sun. This is probably due to its enormous mass and gravity, which may cause it to contract slowly. The amount of contraction needed to explain this extra amount of energy is very small, well below an amount that could be detected by telescopes from Earth. This slow squeezing also means that deep within Jupiter's atmosphere the temperature is many tens of thousands of degrees. A space probe named Galileo is scheduled to explore Jupiter's atmosphere in the mid-1990s.

Jupiter in orbit

JUPITER FACTS
Distance to Sun 483.6 million miles or 5.20 a.u.
Length of year 11.86 Earth years
Orbit eccentricity 0.048
Orbit inclination 1.3 degrees
Diameter 88,732 miles or 11.2 × Earth's
Mass 317.8 × Earth's
Density 1.33
Gravity 2.54 × Earth's
Length of day 9ʰ51ᵐ
Tilt of axis 3.1 degrees

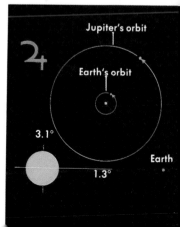

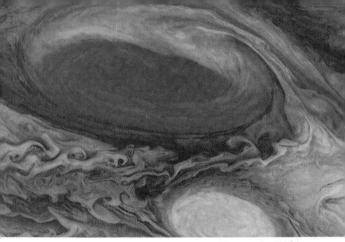

The Great Red Spot on Jupiter, as seen by Voyager 1 NASA

JUPITER'S ATMOSPHERE is its most conspicuous feature. It is about 90 percent hydrogen and 10 percent helium, very close to the proportions of those gases in the Sun. Other elements, such as carbon, sulfur, nitrogen, and oxygen, make up less than 1 percent, but they are important because their chemical compounds are responsible for the colors of the clouds.

The cloud layer is thin, only about 100 miles thick. Methane, ammonia, water vapor, and ice are major components. In color photographs, red clouds, the coolest, are on top; white clouds are below them, slightly hotter. Brown clouds are lower and hotter still, with blue clouds being the lowest and hottest layer. Below the cloud layer is a huge clear atmosphere about 10,000 miles thick. At the base of that, the temperatures and pressures are such that the hydrogen gas is turned into a metal layer 25,000 miles thick. At the very center is probably a rocky core about 13,000 miles in diameter.

The Jovian atmosphere shows bands of clouds, each of which is really a high-speed jet stream of gases. There are some permanent features. One is the Great Red Spot, which seems to be a storm that has lasted 300 years or more. Other colored spots come and go, lasting from days to years.

In the upper atmosphere we can observe lightning

Structure of Jupiter

and auroras, the latter caused by atomic particles hitting the gases in the atmosphere. The planet has an extremely strong magnetic field, and sends out radio waves that can be detected on Earth. It also has very large regions of charged particles around it, similar to, but much more intense than, the Van Allen belts around Earth.

Jupiter has a relationship with its satellite Io that is unique among the planets of the solar system. Strong magnetic fields from Io intersect Jupiter, and charged atomic particles fall into Jupiter's atmosphere from Io's orbit. In addition to causing auroras, these heat the top layers of the atmosphere to more than 1,500 degrees.

Close-up of atmosphere   NASA

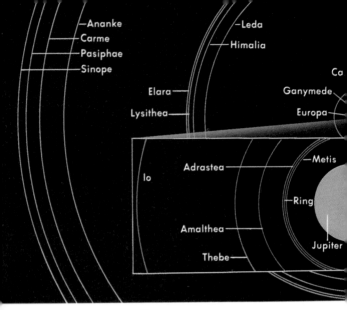

Orbits of Jupiter's satellites

ORBITING JUPITER are a faint ring and a system of satellites, some of which are larger than the smaller planets. Galileo discovered the four largest satellites in 1610.

The large Galilean satellites, which can be considered planets themselves, orbit Jupiter in periods ranging from 1.8 to 17 days. Their orbits are tilted only slightly to the equator of Jupiter, and are almost circular. The smaller satellites are both closer to and more distant from Jupiter than the Galilean satellites. Most are just a few miles across, and have orbits that are both eccentric and inclined to Jupiter. The four outermost satellites revolve about the planet backwards compared to all the others, and are probably captured asteroids. All the satellites are immersed in Jupiter's large, powerful magnetic field.

The ring, about 35,000 miles above Jupiter's cloud tops, is only a few thousand miles broad and less than 20 miles thick. It seems to be made of dark particles about a micrometer in size, and hence is very hard to see. It has been photographed from Earth with special instruments, but the best views have been taken from spacecraft.

SATELLITES OF JUPITER

Name	Year Disc.	Distance from Jupiter (1,000 miles)	Period of Revolution (days)	Mass (× the Moon's)	Density	Diameter (miles)
Metis	1979	79.5	0.294	?	1.97	25(?)
Adrastea	1979	80.1	0.297	?	1.97	16(?)
Amalthea	1892	111.8	0.498	?	1.89	105
Thebe	1979	137.9	0.674	?	1.97	62(?)
Io	1610	262.2	1.769	1.2	1.61	2,256
Europa	1610	416.9	3.551	0.66	1.61	1,951
Ganymede	1610	664.9	7.155	2.02	1.61	3,268
Callisto	1610	1,171.3	16.689	1.46	1.61	2,983
Leda	1974	6,903.4	240	?	1.97	9(?)
Himalia	1904	7,127.1	251	?	1.90	115(?)
Lysithea	1938	7,276.2	260	?	1.93	22(?)
Elara	1905	7,294.9	260	?	1.90	47(?)
Ananke	1951	13,173	631(r)	?	1.95	19(?)
Carme	1938	13,888	692(r)	?	1.93	25(?)
Pasiphae	1908	14,497	735(r)	?	1.90	31(?)
Sinope	1914	14,521	758(r)	?	1.91	22(?)

(r) retrograde motion

Jupiter's ring seen by Voyager   NASA

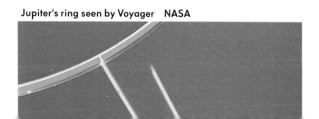

Io photographed by Voyager 1   NASA

IO, Jupiter's innermost Galilean satellite, is one of the strangest and certainly the most active volcanic object in the solar system. Instead of lava, it spews out molten sulfur and sulfur dioxide. Because its surface gravity is so weak, the eruptions jet hundreds of miles out into space and around its globe.

The heat to power the volcanoes probably comes from the competing gravitational tugs of Jupiter and of Europa, the next satellite out, which orbits with a period twice that of Io. These forces squeeze and stretch Io's material, heating much of Io's internal rocks to the melting point. Some heat could also come from interaction with Jupiter's magnetic field. The core of Io is probably solid, silicate material like Earth's crust.

Io is 2,256 miles across, about 100 miles larger than Earth's Moon; it is about 20 percent more massive. Its density leads us to think that Io is made mostly of silicate

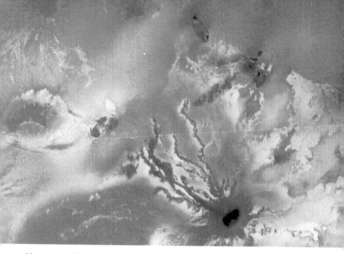

Close-up of Io's surface   NASA

rocks. It orbits 262,000 miles from Jupiter's center, well within the strong magnetic field of the planet. Following Io around its orbit is a huge cloud of gaseous sodium and other elements, tens of thousands of miles long. Io's motion through the magnetic field produces huge electrical currents that strike Io's surface, and also affects the radio waves Jupiter emits.

The surface of Io is reddish, a result of the sulfur from its volcanoes. There are more than 200 volcanic craters on Io, compared with only about 15 on Earth. There are also high mountains, the origin of which is a mystery.

Io and Jupiter's inner satellites are more rocky and hence denser than the outer satellites. This is probably because during the early stages of the solar system the proto-Jupiter was hot and heated its inner satellites, causing the volatile lighter gases to escape. The outer satellites kept their gasses.

Europa   NASA

EUROPA The most puzzling satellite of Jupiter is Europa, second of the Galilean moons. It is a smooth ball 1,951 miles in diameter. In its orbit 417,000 miles from Jupiter's center, Europa takes 3.551 Earth days to circle the planet, exactly twice the time it takes Io.

The lack of any very high surface features is surprising. The photographs taken by the Voyager spacecraft show an intricate maze of several sets of intersecting dark lines. The best guess we have so far is that they are cracks in the surface.

We believe the surface is a layer of water ice that originally welled out of the interior of Europa after its formation. It is probably tens of miles thick, like a frozen ocean covering a rocky core. There are few craters, perhaps because fewer meteoroids struck there or perhaps because the craters get covered over with ice after they form.

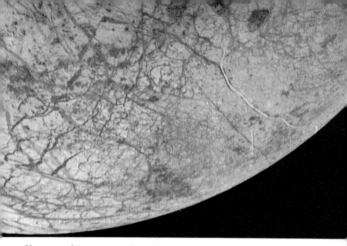

Close-up of Europa's surface   NASA

There are darker and lighter areas on the surface, possibly indicating that there is more than one kind of surface material. It is also possible that the surface is bombarded by streams of atomic particles, like that of Io, and that this may affect the surface colors.

Because the surface seems to be frozen water from the interior of the satellite, there are no high vertical geographical features on Europa. The loftiest features appear to be only a few hundred feet high. There is debate over whether liquid water still exists beneath the icy surface. If so, Europa may be completely surrounded by an ocean with an icy crust. Beneath the crust is probably a rocky core.

Europa and Io are both about the same size and density as Earth's Moon, and hence have about the same gravitational pull. An object of 100 pounds on Earth would weigh only about ⅙ of that on these satellites.

Ganymede   NASA

GANYMEDE This is the most massive and the second largest satellite in the solar system. It is even larger than the planets Mercury and Pluto, but because of its low density, it is only half as massive as Mercury. It is probably composed of a rocky core, covered by a mantle of ice or water, in turn covered by a crust of ice tens of miles thick. Some call it the "ice giant" of the solar system.

In classical Graeco-Roman mythology, Ganymede (sometimes spelled Ganymedes) was an extremely handsome youth taken from Earth to become the servant and cup-bearer to the gods of Olympus. In keeping with astronomical tradition, many of the other satellites also were given names commemorating their mythological connections with Jupiter.

Ganymede is two-thirds of a million miles from Jupiter,

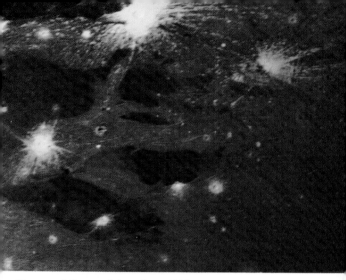

Close-up of Ganymede's surface   NASA

going around it in 7.155 Earth days, just twice the orbital period of Europa, and four times that of Io. When one satellite has a period that is an exact multiple of the period of another satellite, they are said to be in *resonance*.

The surface shows bright and dark areas, craters, extensive grooves, and many-sided blocks of dark material. Two of the dark areas are named for Galileo and for Simon Marius, who discovered the large satellites of Jupiter about the same time Galileo did.

The "blocks" showing on the surface may be chunks of material brought to the surface by motions within the satellite. The grooves are regions, some thousands of miles long, in which the crust spread apart and parts dropped downward. Blocks of crust may be moving, somewhat like the plate tectonics of Earth.

Callisto   NASA

CALLISTO This outermost Galilean satellite is also the most heavily cratered object we know of. Its low density of 1.6 implies that it is largely ice with a rocky core. Orbiting more than a million miles from Jupiter about once every 17 Earth days, Callisto is the second largest of Jupiter's moons, and the third largest moon in the solar system.

Along with Ganymede and Saturn's Titan, Callisto is one of what some planetary scientists describe as a new class of solar system bodies with planetlike sizes but with much lower densities.

This moon is so heavily cratered that the only uncratered portions of the surface are the centers of recent large craters. There are not many high areas because the icy surface is not as strong as rock and cannot support them. Most of the craters are flatter and have lower walls and

Surface features on Callisto   NASA

peaks than do craters on the Moon. Several enormous impact structures are visible, with rings of crushed and broken crust extending hundreds of miles. They resemble some of the larger impact areas on Earth's Moon.

Temperatures on Callisto may get as high as 175 degrees below zero Fahrenheit, but mostly are much colder. At such temperatures, ice behaves like rock, but can flow slowly, as it does in the much warmer glaciers on Earth. Such ice flows in the crust and below it caused cracks and grooves in the surface.

In mythology, Callisto (which means "most beautiful" in Greek) was a favorite nymph of Jupiter. In anger, Jupiter's wife Juno (known in Greek as Hera) turned Callisto into a bear. To save her, Jupiter placed her in the sky as the constellation Ursa Major, the Great Bear.

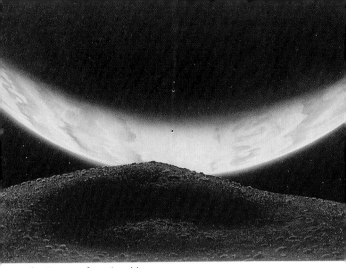

Jupiter seen from Amalthea

JUPITER'S SMALLER SATELLITES Jupiter has more than a dozen small moons that we know about and probably more we have yet to discover.

At half the distance of Io, Amalthea and Thebe orbit Jupiter, taking about half a day. Amalthea, discovered in 1892, is 105 miles across, while Thebe is only about 62 miles across. Closer still to Jupiter are two very small (15- to 25-mile diameter) moonlets very near Jupiter's ring. These "shepherd" satellites help keep the material in the ring together.

Beyond the most distant Galilean satellite, Callisto, 1.17 million miles from Jupiter, there is a big gap with no known satellites. Then, between about 7 and 15 million miles out, there is a group of small bodies that are probably captured asteroids. The largest of them, Himalia, is only 115 miles across. All have quite eccentric and inclined orbits, and not much is known about them. The outer four orbit Jupiter in the opposite direction from all the others.

Because of their small sizes and fairly dark surfaces, very little is known about Jupiter's minor satellites. Only Amalthea and Thebe have been photographed closely; the others have so far been seen only as tiny points of light. The Galileo spaceprobe may provide us with closer views of a few more of these small satellites in the mid-1990s.

Two curious groups of small objects might be called satellites of both Jupiter and the Sun. They are the *Trojan asteroids*, which orbit the Sun in the same orbit as Jupiter. They are near the *Lagrangian Points* (named after the French astronomer who first found that such objects could have stable orbits) in Jupiter's orbit—that is, at the points 60 degrees ahead of and behind the giant planet, making them almost 500 million miles from Jupiter. They are called Trojans because all of these asteroids are named for characters from the Trojan War. Most of the ones preceding Jupiter in orbit are named for Greek warriors, and those following Jupiter have names of Trojan warriors; however, there is one Trojan name in the Greek group and one Greek in with the Trojan group. There are slightly over a dozen bodies known, and there probably are several dozen more too small to be seen from Earth. They may be escaped satellites of Jupiter, or asteroids captured into this particular orbit by Jupiter's strong gravity.

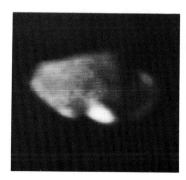

Amalthea   NASA

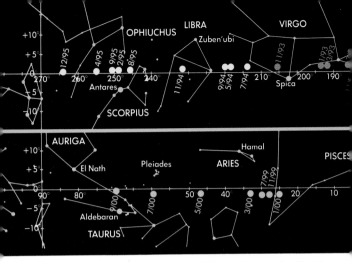

OBSERVING JUPITER Jupiter's huge size and reflective clouds make it brighter than anything except the Sun, Moon, or Venus. Orbiting the Sun once every 11.86 Earth years, it repeats its positions relative to Earth every 399 days. Near times of opposition, it can come within 360 million miles of Earth. The chart above shows Jupiter's changing position against the background of the stars of the zodiac during this decade.

The larger satellites seem to perform a cosmic ballet around Jupiter, changing relative positions night to night. This makes for interesting viewing through binoculars or a small telescope. Positions may be found in the *Observer's Handbook* and astronomy magazines (see p. 157).

A small telescope will also show the banded atmosphere of the planet, and its major features, including the famous Great Red Spot. Some of the smaller satellites can also be seen, but the smallest ones require a large telescope, or even a spacecraft, to be seen. The ring is not visible in small telescopes.

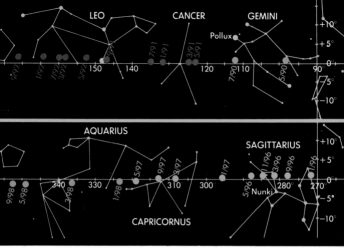

Finding chart for Jupiter

Jupiter sometimes approaches other planets in the sky as seen from Earth. On August 12, 1990, Jupiter and Venus will pass only 0.04 degrees apart in the sky (for comparison, the Moon is 0.5 degrees across) and about 21 degrees west of the Sun. They will be only 0.01 degrees apart on May 17, 2000, but more difficult to see since they will be only 7 degrees west of the Sun.

BEST TIMES FOR VIEWING JUPITER

Date of Opposition	Distance from Earth (miles)
1991 Jan 29	400 million
1992 Feb 29	410 million
1993 Mar 30	414 million
1994 Apr 30	412 million
1995 Jun 1	402 million
1996 Jul 4	389 million
1997 Aug 9	376 million
1998 Sep 16	368 million
1999 Oct 23	368 million
2000 Nov 28	376 million

Note The best times for viewing Jupiter are the several months around the dates of opposition.

A view of Saturn

SATURN

Probably the most spectacular planet, Saturn marked the outer boundary of the solar system to the pre-telescopic astronomers. Its stately motion around the Sun, taking 29.46 Earth years, led to it being associated with time, as its ancient Greek name, Chronos, shows. Saturn was also associated with agriculture, and its astronomical symbol is a stylized scythe.

Being farther from the Sun, Saturn is colder, less bright, and less colorful than Jupiter. Yet it sports a magnificent set of rings that were discovered by Galileo in 1610.

Saturn is probably unique among the planets in having an average density of only 0.7. This means that if you could find a body of water big enough to hold it, Saturn would float. (This may also be true for Pluto.) Average *density* is simply mass divided by volume. A planet's mass can be measured by studying the planet's gravitational force on its satellites. Its volume is determined by its size and shape.

Saturn has a magnetic field—about 550 times stronger than that of Earth, but only about 3 percent that of Jupiter—which holds charged atomic particles in vast belts around the planet, similar to Earth's Van Allen belts. This *magnetosphere* lies outside the A ring (see p. 118). Many of the atomic particles in these belts come from Saturn's larger inner satellites, particularly Titan. Like Jupiter's but much weaker, Saturn's magnetosphere sends out faint radio waves. The strength of the waves changes slightly over time, and seems to be affected by the motion of some of the satellites, particularly Dione.

Three centuries of telescopic examination were followed by three space probes, providing us with the first close-up views of the planet and its moons. Pioneer 11 in 1979, followed by Voyager 1 in 1980, and Voyager 2 in 1981 discovered several new moons, and found that there are really thousands of small "ringlets" making up the few major rings that we see from Earth.

SATURN FACTS

Distance to Sun 886.7 million miles or 9.539 a.u.
Length of year 29.46 Earth years
Orbit eccentricity 0.056
Orbit inclination 2.5 degrees
Diameter 74,565 miles or 9.41 × Earth's
Mass 95.16 × Earth's
Density 0.70
Gravity 1.08 × Earth's
Length of day 10^{h}39^m
Tilt of axis 26.7 degrees

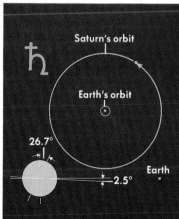

Saturn in orbit

Saturn and its rings   NASA

THE COMPOSITION OF SATURN is much like that of Jupiter, mostly hydrogen and helium. Saturn's smaller mass allows a more extended atmosphere than Jupiter's, and cooler temperatures make the chemical reactions in the atmosphere less colorful. The two principal minor gasses found in Saturn's atmosphere are methane and ammonia.

The planet is 95 times as massive as Earth, about 30 percent the mass of Jupiter. Because of its low density, a 100-pound person on Earth would weigh 108 pounds on Saturn. It spins on its axis rapidly, one day taking 10.66 Earth hours. Because of this speed, Saturn is flattened at the poles more than any other planet. This oblateness is noticeable through a small telescope or in photographs.

Like Jupiter, all we see are the tops of clouds of methane

and ammonia. The highest clouds are pinkish in color, composed mostly of ammonia ice crystals at a temperature of −170 degrees Fahrenheit. Below these are brownish clouds made of crystals of a chemical called ammonium hydrosulfide, at a temperature of about 10 degrees. The lowest cloud layer, bluish in color, is composed of water vapor and ice, at a temperature of about 50 degrees.

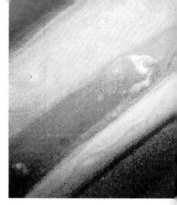

Close-up of atmosphere
NASA

Below the clouds is a clear atmosphere even deeper than Jupiter's, with pressure increasing as you descend. When the pressure reaches 3 million times Earth's atmospheric pressure, the hydrogen becomes metallic.

At the center of the planet is a core of ice and silicate rocks about two and a half times the size of Earth. Also like Jupiter, Saturn gives off more energy than it gets from the Sun, again due to contraction and also perhaps to a process not yet occurring on Jupiter—the condensation of atmospheric helium into helium "raindrops" that fall downward, releasing heat.

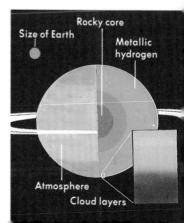

Internal structure of Saturn

Size of Earth

Rocky core

Metallic hydrogen

Atmosphere

Cloud layers

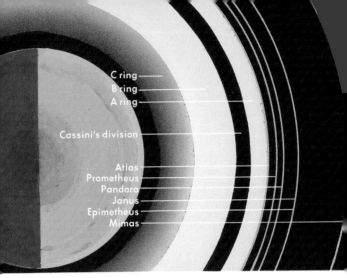

C ring
B ring
A ring
Cassini's division
Atlas
Prometheus
Pandora
Janus
Epimetheus
Mimas

Saturn's rings

THE RINGS OF SATURN Discovered in 1610 by Galileo, the rings of Saturn were thought to be unique in the solar system until recently. While Saturn's rings are the biggest and most spectacular, we now know that Jupiter, Uranus, and Neptune have ring systems, too.

Galileo originally thought that Saturn was three close objects. The poor image in his telescope also made him think that the planet had "ears" or "handles." It was not until 1659 that Christian Huygens correctly explained the rings of Saturn. Even then, it was not known whether the rings were solid sheets of material or if they were made up of small particles.

Today we know that the rings are indeed a multitude of small particles, mostly ice or ice-covered rock, ranging from a few inches to tens of feet in diameter, spaced feet

Voyager's view of the rings   NASA

apart. They occupy a very thin disk only around 100 feet thick, but stretching more than a quarter of a million miles from the inner edge of the innermost ring to the far edge of the outer ring. The ring material is so thin that often distant stars and Saturn itself may be seen through the rings.

The inner ring, called the D ring or "crepe" ring because it is so thin and transparent, is only 4,000 miles above Saturn's equator. Next outward is the C ring, then the B ring. Between this and the A ring is a gap where almost no particles orbit, called *Cassini's division*, after the Italian-French astronomer Jean Dominique Cassini who discovered it in 1675. (The gap is similar to the Kirkwood gaps, p. 93, in the asteroid belt, and occurs where a particle would have an orbital period of half that of the major satellite

Rings showing "spokes"   NASA

Mimas.) Within the A ring is a smaller but similar gap, *Encke's division*, named for German astronomer Johann Encke.

Space probes have discovered three rings beyond the A ring: the F ring, the G ring, and finally the E ring. The outermost edge of the E ring may extend as far as 300,000 miles from Saturn.

Thousands of "ringlets" make up the rings. Some are clumpy, some seem "braided" together. There are also radial features that look like spokes in the rings. Some of the rings are not perfectly circular. Just how and what keeps the rings together is a very complex dynamical problem not yet completely solved. The narrowest ringlets are as small as 50 feet across. The widest ring is more than 12,000 miles across.

Each particle in the rings obeys the same laws of motion as planets do. The particles closer to Saturn orbit faster than the ones farther out. The particles' motion is disturbed

Back-lit view of the rings   NASA

by small moonlets orbiting within and on the edges of some of the rings. These "shepherd" satellites help keep the rings from flying apart.

Saturn's thin rings orbit in the plane of the planet's equator, which is tilted 30 degrees to the plane of its orbit. Every 15 years, half of Saturn's 30-year orbital period, Earth passes through the plane of the rings. This occurs several times spaced a few months apart. At such times, the rings are edge-on to us, and hence invisible for many months around those times (see pp. 130-131). Seven and a half years after such times we see the rings widely tilted, and thus very visible.

Saturn's magnetic field extends into the rings, and it is possible that charged bits of ice, caused by collisions between ring particles, may be carried down into the top of the atmosphere and cause a kind of "cloud seeding" that makes atmospheric haze condense and drop to lower levels.

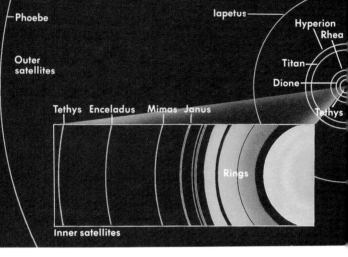

Orbits of Saturn's satellites

THE SATELLITES OF SATURN can be divided into three groups: (1) very small moonlets, a few tens of miles in size; (2) satellites in the 50- to 1,000-mile diameter range; (3) Titan.

THE SMALLEST SATELLITES orbit near the rings, and help keep them in place. Typically, their orbits are in the plane of Saturn's equator and are almost perfectly circular. Most of them are nonspherical in shape, more like eggs or footballs. We know little about their properties.

THE MEDIUM-SIZED SATELLITES typically have slight orbital inclinations and eccentricities. The smallest of them is outermost Phoebe, more than 8,000,000 miles from Saturn. Discovered in 1898, it travels backwards around Saturn, the first retrograde moon ever found. It may be a captured asteroid.

THE LARGEST satellite other than Titan (see pp. 128-129) is Rhea, 475 miles across, orbiting 327,000 miles from Saturn. Rhea shows extensive fractures on its surface, and possibly snow. Hyperion, orbiting 920,000 miles from Saturn, is the largest nonspherical moon.

Rhea   NASA

SATELLITES OF SATURN

Name	Year Disc.	Distance from Saturn (1,000 miles)	Period of Revolution (days)	Mass (× the Moon's)	Density	Diameter (miles)
1990 S18	1990	82.9	0.58	?	?	12
Atlas	1980	85.1	0.601	?	?	19
Prometheus	1980	86.3	0.613	?	?	62
Pandora	1980	88.2	0.628	?	?	56
Epimetheus	1966	93.8	0.695	?	?	76
Janus	1966	93.8	0.695	?	?	118
Mimas	1789	116.2	0.942	.0005	1.2	242
Enceladus	1789	147.9	1.370	.001	1.1	311
Tethys	1684	183.3	1.888	.01	1.2	659
Telesto	1980	183.3	1.888	?	?	16
Calypso	1980	183.3	1.888	?	?	16
Dione	1684	234.9	2.737	.014	1.4	696
Helene	1980	234.9	2.737	?	?	19
Rhea	1672	326.8	4.517	.033	1.3	951
Titan	1655	758.7	15.945	1.83	1.88	3,449
Hyperion	1848	920.3	21.276	?	?	158
Iapetus	1671	2,213	79.331	.026	1.2	907
Phoebe	1898	8,053	550(r)	?	?	137

(r) = retrograde motion

Several of the smaller satellites are irregular in shape.

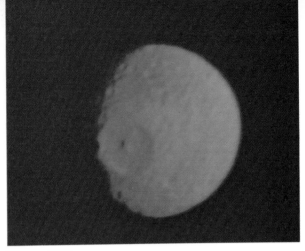

Mimas   NASA

MIMAS, Saturn's innermost major satellite, orbits 116,000 miles from the planet, once every 0.942 Earth days. The moon is spherical, 242 miles in diameter, with a density of only 1.2.

Like almost all the Saturnian satellites, its surface is made mostly of ice and is heavily cratered. At a temperature almost 300 degrees below zero, ice is like rock in its properties. The surface resembles the cratered highlands of the Moon, except that the ice makes it much more reflective. The largest crater, named Herschel, is 84 miles across, and the collision that produced it probably almost destroyed Mimas long ago.

Mimas maintains a fixed orientation as it circles Saturn, with one side always facing the planet. The Voyager spacecraft mapped much of Mimas. Other than the largest crater, Herschel, most of the other features on this satellite are named for characters from Britain's Arthurian legends, including craters called Merlin, Arthur, and Galahad.

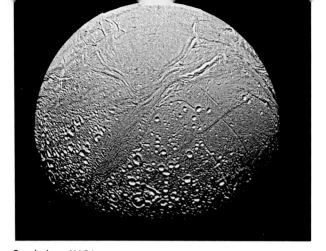

Enceladus   NASA

ENCELADUS is also a satellite made of ice. It is 311 miles in diameter and orbits 148,000 miles from Saturn. It, too, is cratered, but also shows what appears to be the results of vulcanism—not of lava, but of water. There are also long, narrow valleys. Astronomers deduce continuing vulcanism because there are no craters larger than about 20 miles across, and all the craters are younger than those on other satellites. Something must erode them or fill them in. Also, the surface is more reflective than that of the other moons, meaning it must be newer material. Probably the interior of Enceladus is still liquid and still in motion. The source of this internal heat which keeps Enceladus active is unknown.

Like many of the other smaller satellites, Enceladus keeps one face toward Saturn at all times. The Voyager 2 space probe flew fairly close to this satellite and mapped part of its surface in better detail than most of the other Saturnian satellites.

Tethys   NASA

TETHYS orbits Saturn 183,000 miles from it, once every 1.888 Earth days. An interesting feature is that there are two "co-orbital" moonlets, named Telesto and Calypso, orbiting in exactly the same orbit as Tethys. Only 16 miles in size, they lie 60 degrees ahead of and behind Tethys, similar to the way the Trojan asteroids precede and follow Jupiter in its orbit. (The more-distant satellite Dione also has an orbital companion, Helene.) Their existence is not fully understood.

This 659-mile diameter icy ball is cratered, but some parts show more craters than others. This leads us to think that internal geological processes have smoothed some regions of the satellite. Tethys also shows long branching valleys, which indicate that this satellite is still geologically active. A few of the features on Tethys, mapped by Voyager spacecraft, have been named after figures in Greek mythology.

Iapetus   NASA

IAPETUS, 2,213,000 miles from Saturn, is one of the most bizarre objects we know of in the solar system. It is 907 miles in diameter, and orbits Saturn once every 79 Earth days. Its orbit is the most highly inclined of all Saturn's satellites except for Phoebe's, which is tilted 15 degrees from the plane of the planet's equator and rings.

Like most of Saturn's satellites, it always keeps one face toward the planet. The curious feature is that its trailing half is covered with bright icy material, while its forward half is made of some much darker material. The moon has a low density of only 1.2, which means it is mostly ice. Somehow the leading edge picked up a dark coating of unknown material.

The leading, dark hemisphere reflects less than 5 percent of sunlight hitting it. We don't yet know if the dark material came from outside the satellite or from within the satellite itself.

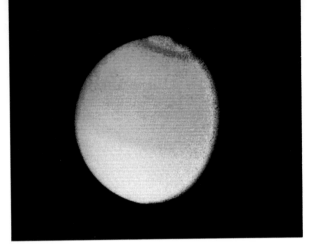

Titan   NASA

TITAN is the largest satellite in the solar system. It is 3,449 miles across, larger than the planets Mercury and Pluto. Once every 16 Earth days Titan orbits Saturn at a distance of 759,000 miles.

Titan's other unique feature is its thick atmosphere. Titan's is even denser than Earth's at the surface. It is also the only atmosphere in the solar system, other than our own, that is mainly nitrogen—about 80 percent. There is no oxygen. This atmosphere gives Titan a faint orange color.

Despite a surface temperature no warmer than 300 degrees below zero Fahrenheit, the atmosphere is active. In addition to nitrogen, methane and argon make up most of the atmosphere. Minor gases include hydrogen cyanide, acetylene, and other hydrocarbons. They produce a thick chemical smog layer more than 120 miles thick, and hence we have never seen Titan's surface. There are also methane clouds and a higher haze layer. If you were on the surface, you could probably see for only a few tens of miles. The atmospheric pressure there is 1.6 times that at the surface of Earth.

There may be oceans of methane on the surface, perhaps covering it completely. The surface is ice. Titan may contain a huge reservoir of hydrocarbon chemicals, the richest in the solar system, produced by reactions among its atmospheric gases raining down to the surface.

Titan is so large that after it formed, it differentiated. That is, internal heat and heat from the impacts of meteoroids caused the heavier materials to form a rocky core, covered by a liquid mantle and an icy crust. By now the mantle has probably cooled off and solidified.

When Voyager photographed Titan, we noticed that the northern hemisphere was somewhat darker than the southern hemisphere. This may indicate that Titan experiences seasons. If so, each one would last 7.5 years, since it takes Saturn 30 years to orbit the Sun.

Surface of Titan

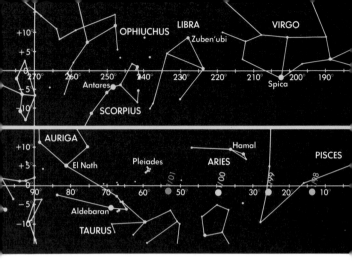

OBSERVING SATURN When nearest Earth, around times of opposition, Saturn is among the brighter objects in the sky. Near conjunction, however, it is much fainter. It often appears slightly yellowish.

Saturn's synodic period is just two weeks longer than a year, and it takes 30 years for it to circle the zodiac. At opposition it comes within about 750 million miles of Earth. The chart above shows the position of Saturn against the background of stars on January 1 of each year of this decade.

In the middle of the decade of the 1990s, Earth passes through the ring plane on May 21, 1995, August 11, 1995, and again on February 11, 1996. Saturn is then tilted in such a way that its rings appear edge-on to us, and are therefore invisible because they are so thin. Thus the rings will be difficult to view throughout the middle of the decade.

Saturn's rings and a few of its larger satellites are visible

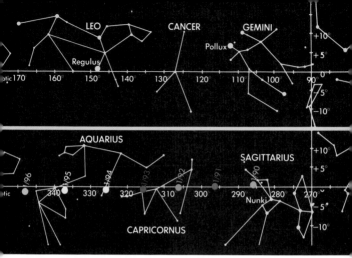

Finding chart for Saturn

in small telescopes, but because it is so far away little surface detail can be seen on Saturn except for faint atmospheric bands and the major divisions of the rings.

BEST TIMES FOR VIEWING SATURN

Date of Opposition		Distance from Earth (miles)
1990	Jul 14	835 million
1991	Jul 27	831 million
1992	Aug 7	825 million
1993	Aug 19	817 million
1994	Sep 1	807 million
1995	Sep 14	799 million
1996	Sep 26	789 million
1997	Oct 10	779 million
1998	Oct 23	771 million
1999	Nov 6	763 million
2000	Nov 19	756 million

Note The best times for viewing Saturn occur for several months around the dates of opposition.

URANUS

Uranus, the third-largest planet, was the first one discovered since ancient times. English astronomer Sir William Herschel found it by chance while searching the sky with a telescope on March 13, 1781. Herschel named it "Georgian," after his patron, King George III of England. Others called it "Herschel." The German astronomer Johann Bode suggested it should have a mythological name like the other planets, and proposed the name "Uranus," after the very ancient Greek deity of the heavens. (The proper modern pronunciation is "Yoo-ray-nus.") Uranus is so far from the Sun that its discovery doubled the size of the known solar system. Its orbit marked the solar system's outer boundary until 1846, when Neptune was discovered.

Until the mid-1980s, Uranus was thought to have only five satellites: two were discovered by Herschel in 1787, two more were found in 1851, and another was sighted in

A view of Uranus

1948. In 1986 Voyager 2 flew past the planet and found ten smaller moons. Voyager also confirmed the Earth-based observation that Uranus has a very faint set of rings.

Uranus is 14.5 times the mass of Earth, but has a density only slightly greater than water. A 100-pound person would weigh 91 pounds there. It takes Uranus 84 Earth years to circle the Sun once, at an average distance of 1.8 billion miles. Sunlight takes more than two and a half hours to reach this planet.

Uranus is usually grouped with Jupiter and Saturn as a gas-giant, or Jovian, planet, as is more distant Neptune. Some planetary scientists now think that Uranus and Neptune—both about 32,000 miles in diameter—should be considered a separate class of planet on their own. These two planets are only about a third of the size of Jupiter and about five percent of its mass. They have slightly higher densities. Being farther from the Sun, they are colder. Whereas Jupiter and Saturn are mostly hydrogen and helium, calculations predict that Uranus and Neptune have substantial quantities of heavier elements such as oxygen, nitrogen, silicon, and iron.

Uranus in orbit

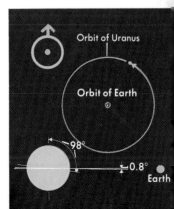

URANUS FACTS

Distance to Sun 1,783,000,000 miles or 19.182 a.u.
Length of year 84.01 Earth years
Orbit eccentricity 0.047
Orbit inclination 0.8 degrees
Diameter 31,814 miles or 4 × Earth's
Mass 14.500 × Earth's
Density 1.30
Gravity 0.91 × Earth's
Length of day 17^{h}14^m
Tilt of axis 98 degrees

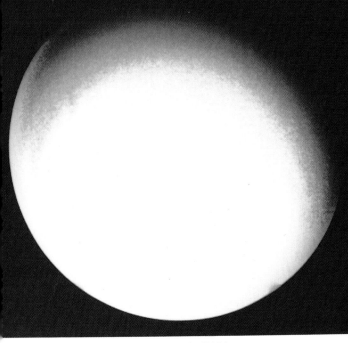

Uranus photographed by Voyager 2   NASA

URANUS seen from Earth shows a bland surface, really the tops of clouds in its hydrogen-helium atmosphere. Much colder than Jupiter or Saturn, there is very little color and few details visible in the atmosphere. The planet appears slightly blue-green.

Though it may look bland, Uranus is a peculiar planet in several ways. The other planets (with the exception of Pluto) have their north and south poles (the axis of their spin, or rotation) tilted over less than 30 degrees measured from a line perpendicular to their orbital planes. (Earth is tilted 23.4 degrees, and this causes the seasons we experience; see p. 64.) Uranus is tilted over almost 98 degrees! This means that its north pole points almost along the plane of its orbit; actually it points slightly to the south side of the solar system. Thus there are times during Uranus' orbit when its north pole points almost directly at the Sun; one quarter orbit (21 years) later the Sun would lie over Uranus' equator; and 21 years after that the south pole is pointed at the Sun. This must make for very strange seasons on Uranus.

Because its satellites orbit in the plane of Uranus' equator, this tilt also means that the satellites move almost perpendicular to the plane of Uranus' orbit. The only other planet in the solar system with a tilt like this is Pluto.

Uranus also has a magnetic field, like many of the other planets. But unlike the other planets—which have their north magnetic poles near their north geographic poles—the poles of Uranus' magnetic field are tilted over more than 60 degrees compared to its axis of spin. Furthermore, the center of the magnetic field is not at the center of the planet.

After hydrogen and helium, the main component of the atmosphere is methane. The only atmospheric features that are visible are bands of polar haze produced by the action of sunlight on methane, although Voyager did observe auroras high in the atmosphere.

Structure of Uranus

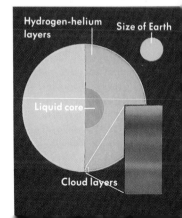

Hydrogen-helium layers

Size of Earth

Liquid core

Cloud layers

THE SATELLITES OF URANUS total 15 in number but fall into two different-sized groups. The five outer ones are several hundred miles in size, and are major worlds like the satellites of Jupiter and Saturn. Ten smaller moons, all orbiting closer, were discovered by Voyager in 1986. All the satellites and the rings are in the highly-tilted equatorial plane of Uranus.

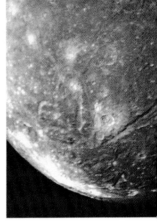

Ariel   NASA

Each of the major moons has its own peculiarities. Some are very complex geologically and show evidence of active geology.

The inner small satellites, some of which act as "shepherds" for some of the rings, are all very dark, as dark as charcoal, like the ring material itself.

The rings of Uranus   NASA

Before Voyager we knew of the existence of several rings of small dark particles, orbiting nearer to Uranus than Miranda. Voyager found 2 more, for a total of 11, plus unexplained segments, or "arcs," of rings. There are now 11 rings known, but their origin and what keeps them in place are still debated.

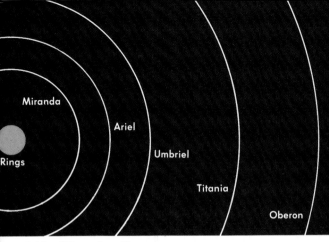

Orbits of Uranus' satellites

SATELLITES OF URANUS

Name	Year Disc.	Distance from Uranus (1,000 miles)	Period of Revolution (days)	Mass (× the Moon's)	Density	Diameter (miles)
Cordelia	1986	30.944	0.333	?	?	16
Ophelia	1986	33.430	0.375	?	?	19
Bianca	1986	36.785	0.433	?	?	25
Cressida	1986	38.401	0.463	?	?	37
Desdemona	1986	38.960	0.475	?	?	34
Juliet	1986	40.016	0.492	?	?	53
Portia	1986	41.072	0.513	?	?	68
Rosalind	1986	43.496	0.558	?	?	34
Belinda	1986	46.789	0.621	?	?	40
Puck	1986	53.438	0.763	?	?	96
Miranda	1948	80.716	1.41	.001	1.26	301
Ariel	1851	118.620	2.521	.018	1.65	721
Umbriel	1851	165.285	4.146	.017	1.44	739
Titania	1787	271.104	8.704	.047	1.59	1,000
Oberon	1787	362.508	13.463	.040	1.50	963

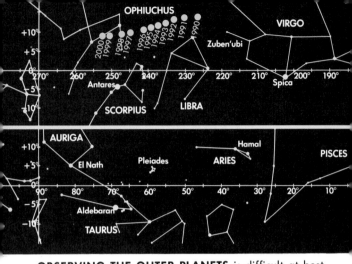

OBSERVING THE OUTER PLANETS is difficult at best. The chart above shows the positions of the outer planets against the background of the zodiac for January 1 of each year of this decade. Uranus is only visible with the unaided eye under perfect conditions, but it is visible through binoculars or a small telescope if you know just where to look. Neptune can be seen with a small telescope, and, like Uranus, appears only as a small greenish disk. Pluto requires a large telescope to be seen.

Uranus' satellites Titania and Oberon are visible in small instruments; the other satellites are too faint. The rings are not visible even in the largest telescopes, and can be detected from Earth only by special instruments. Neptune's satellite Triton is also visible in amateur telescopes. Pluto's Charon requires a large telescope.

Because they are so far away to begin with, these planets vary less in brightness as seen from Earth than do closer planets. Their synodic periods are all just a few days more than one year.

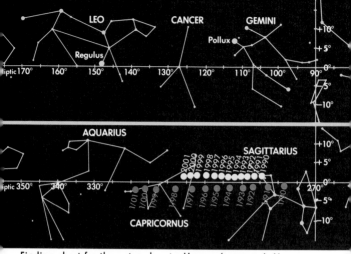

Finding chart for the outer planets: Uranus (magenta), Neptune (yellow), and Pluto (blue)

BEST TIMES FOR VIEWING URANUS

Date of Opposition		Distance from Earth (miles)
1991	Jul 4	1.716 billion
1992	Jul 7	1.721 billion
1993	Jul 12	1.727 billion
1994	Jul 17	1.733 billion
1995	Jul 21	1.738 billion
1996	Jul 25	1.743 billion
1997	Jul 29	1.748 billion
1998	Aug 3	1.752 billion
1999	Aug 7	1.756 billion
2000	Aug 22	1.759 billion

BEST TIMES FOR VIEWING NEPTUNE All oppositions in the 1990s occur in July, with a typical Earth-Neptune distance at opposition of 2.71 billion miles.

BEST TIMES FOR VIEWING PLUTO All oppositions in the 1990s occur in mid-May, with a typical Earth-Pluto distance at opposition of 2.68 billion miles. Because of the eccentricity of its orbit, Pluto will be closer to Earth than Neptune until 1999.

Neptune as seen from the surface of Triton

NEPTUNE

Following the discovery of Uranus, astronomers noted it did not follow exactly the path Newton's and Kepler's Laws predicted. The English astronomer J. C. Adams and the French scientist J. J. Leverrier independently proposed that a still more distant planet was altering its motion, and they predicted where that new planet would be seen. In 1846, the German Johann Bode found the planet where they predicted, again almost doubling the size of the solar system. This planet in the depths of space was named for the Roman god of the depths of the sea. Its symbol is a trident.

The fourth-largest planet, Neptune takes 164.79 Earth years to circle the Sun, at an average distance of 2.793 billion miles. Its orbit is almost circular, inclined less than 2 degrees. We know that Neptune is 17.2 times as massive as Earth, with an average density of 1.76. A 100-pound person would weigh 119 pounds on Neptune.

The combination of Neptune's almost circular orbit and Pluto's highly eccentric orbit (see p. 148) means that Pluto is for a few years closer to the Sun than Neptune. This unusual situation is occurring now, and will last until 1999. It won't occur again for more than two centuries.

Until recently not much was known about Neptune since it could be studied only by large telescopes from Earth. In August, 1989, the Voyager 2 spacecraft made its final planetary encounter, revolutionizing our knowledge of this outermost of the gas-giant planets. Voyager found six new satellites, adding to the two discovered from Earth, and a faint series of rings around the planet.

NEPTUNE FACTS

Distance to Sun 2.793 billion miles or 30.058 a.u.
Length of year 164.79 Earth years
Orbit eccentricity 0.009
Orbit inclination 1.8 degrees
Diameter 30,198 miles or 3.81 × Earth's
Mass 17.204 × Earth's
Density 1.76
Gravity 1.19 × Earth's
Length of day 16.03 hours
Tilt of axis 29.6 degrees

Neptune in orbit

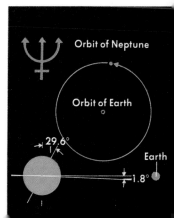

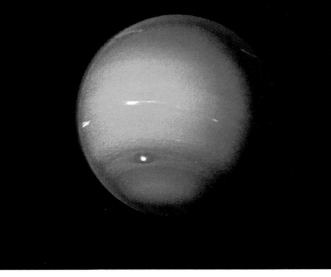

Neptune as seen by Voyager 2   NASA

NEPTUNE'S ATMOSPHERE is at least as active as Uranus' atmosphere, and much more colorful—which is surprising, since Neptune is farther from the Sun than Uranus, and hence colder. Sunlight at Neptune is less than five percent as strong as at Jupiter's orbit. Voyager detected very high wind speeds, over 700 miles an hour, blowing westward— even though Neptune turns eastward. This had led astronomers earlier to think that the planet rotated more slowly than the 16 hours and 3 minutes it takes to turn once. The earlier estimate of a little over 17 hours is the period for the clouds at the top of the atmosphere.

Several dark areas are visible floating high in Neptune's atmosphere, and they change over a period of days. The largest feature, called the "Great Dark Spot," is moving at about 760 miles an hour. Auroras were detected high in the atmosphere, not just in the polar regions as expected, but all over the planet—for reasons not yet understood. Like the other gas-giant planets, the atmosphere is mostly hydrogen and helium, but most of the visible features are a result of methane clouds. It appears that the Sun's ultraviolet light converts methane high in the atmosphere into hydrocarbons that then drop toward the surface. When they reach lower, hotter layers they evaporate back into methane again, which then rises to the top of the atmosphere.

There is evidence of global atmospheric motion, whereby heat from some latitudes is carried to polar regions.

Neptune has a magnetic field inclined to its rotation axis by 50 degrees. The center of the magnetic field is offset by almost half Neptune's radius from the center of the planet. Why this is so is not yet understood. This is similar to the situation for Uranus, and means that the magnetic field probably arose during the planet's origins, not by some unique accident. The presence of a magnetic field means that Neptune must contain a region of electrical conducting fluid that is in motion.

Like the other Jovian planets, Neptune emits more energy than it gets from the Sun. In Neptune's case, it emits 2.7 times the solar energy. This may account for the fact that its atmosphere is more active and colorful than that of Uranus.

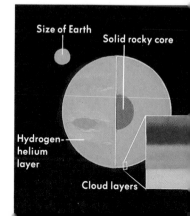

Neptune's structure

Size of Earth

Solid rocky core

Hydrogen–helium layer

Cloud layers

Triton   NASA

NEPTUNE'S SATELLITES AND RINGS When Neptune was discovered in 1846, its largest satellite, Triton, was also found. A century later, in 1949, Nereid was discovered orbiting at five and a half million miles from Neptune, taking a year to orbit once. Because of Neptune's great distance, no more satellites were found by terrestrial observers. In 1989, during Voyager's last planetary encounter, six new satellites were seen. One of them is larger than Nereid but had not previously been seen because it is close to Neptune and thus lost in the planet's glare as viewed from Earth.

Neptune's rings   NASA

Voyager also found several faint rings of dust and small rocks circling the planet. The "Main Ring" is clumpy, and lies 39,000 miles from Neptune's center. At 33,000 miles from Neptune is the "Inner Ring," and closer in at 26,000 miles is the "Inside Diffuse Ring." Just outside the inner ring is a flat sheet of material called the "plateau." These are all much less dense than Saturn's rings, and not visible from Earth.

The greatest surprise of Voyager's visit was Triton. It has a very thin, mostly nitrogen atmosphere, about one-thousandth the

The satellite 1989 N1   NASA

pressure on Earth. The surface shows extensive cratering and scarring. No other body in the solar system combines all the types of features Triton shows. Its density is about twice that of water, meaning that it contains a higher proportion of rock than other icy satellites. Dark and light surface material indicates that some sort of vulcanism may still be going on, but the "lava" is not molten rock; it is probably liquid nitrogen. Planetary scientists think Triton, though actually slightly larger than Pluto, may in fact be a lot like that planet.

SATELLITES OF NEPTUNE

Name	Year Disc.	Distance from Neptune (1,000 miles)	Period of Revolution (days)	Mass (× the Moon's)	Density	Diameter (miles)
1989 N6*	1989	29.9	0.296	?	?	31
1989 N5*	1989	31.1	0.313	?	?	56
1989 N3*	1989	32.6	0.333	?	?	87
1989 N4*	1989	38.5	0.396	?	?	100
1989 N2*	1989	45.7	0.554	?	?	124
1989 N1*	1989	73.0	1.121	?	?	261
Triton	1846	220	5.877	0.29	?	1,689
Nereid	1949	3,478	365.21	?	?	186

*These are provisional names until permanent names are adopted.

Pluto's motion among the stars   PALOMAR

PLUTO

When it turned out that Neptune did not account for all the disturbances in Uranus' orbit (p. 140), astronomers W. H. Pickering and Percival Lowell predicted that another more distant planet must exist beyond Neptune. Lowell himself founded an observatory and funded a search he did not live to see completed.

PLUTO WAS DISCOVERED in 1930. However, it is so small that it cannot account for the motions of Uranus' orbit. The predictions were spurious, and Pluto's discovery was a fortunate accident due to the careful sky survey at the Lowell Observatory by American astronomer Clyde Tombaugh. The new planet was given the name Pluto, after the Roman god of the underworld. The first two letters of the name are also Lowell's initials, and its symbol is a combination of P and L.

More than 3.6 billion miles out, Pluto is small and icy, and some astronomers have even proposed "demoting" it from the status of a planet to that of an asteroid. It may be an escaped satellite of Neptune. The Sun seen from Pluto is just a very bright star, about 1,000 times brighter than the full moon is on Earth. Sunlight takes 5½ hours to reach Pluto.

Because it is so far away and dim, not much is known with certainty about Pluto, and no space probes are now scheduled to visit this the most remote of the planets.

PLUTO FACTS

Distance to Sun 3.666 billion miles or 39.439 a.u.

Length of year 247.69 Earth years

Orbit eccentricity 0.250

Orbit inclination 17.2 degrees

Mass 0.0026 (?) × Earth's

Diameter 1,429 miles or 18 × Earth's

Density 1.1 (?)

Gravity 0.05 (?) × Earth's

Length of day 6.3867 Earth days

Tilt of axis 94 degrees

Pluto and Charon in orbit

PLUTO AND CHARON Pluto is thought to be about 1,429 miles in diameter, mostly ice, and covered with frozen methane. Its density is thought to be just slightly greater than that of water. Its orbit is highly eccentric, taking it closer to the Sun than Neptune at times. Its 248-year orbit took it closest to the Sun in 1989, and until 1999 Pluto will be closer to the Sun than Neptune. Pluto is appropriately named for the god of the underworld because it orbits far from the Sun in a perpetual twilight.

Pluto appears only as a speck of light even in the largest telescopes. Much of the information about it is uncertain. Before 1978 we thought Pluto was larger than we now know it to be, and it was thought to be the second-smallest planet, after Mercury. This was because our photographs blurred together the image of Pluto itself and the image of its satellite. Since we could see no surface features, it was impossible to know how long it took to rotate on its axis.

In 1978 a satellite was discovered and named Charon. It is so close to Pluto that we believe they are gravitationally locked together. Thus Pluto rotates in the same period of time as Charon revolves about it, a little over six days. Pluto's axis seems tilted over (by 94 degrees), much like that of Uranus. Pluto and Charon both keep the same face toward each other. Charon, 808 miles across, is the largest satellite compared to its primary planet in the solar system. It is over half the size of Pluto. Charon is named for the mythological figure who ferried the dead across the River Styx into the underworld.

Careful measurements by telescopes on Earth have revealed signs of what appears to be frozen methane on the surface of Pluto and its satellite. Their surfaces are probably entirely covered with ice, and are only about 10,000 miles apart where they face each other.

Pluto's mass is thought to be only 0.003 that of Earth's, so that a 100-pound person on Earth would weigh only 5 pounds on Pluto. The temperature on the surface is 350 degrees below zero Fahrenheit, and it probably has a very thin atmosphere of methane, 1/60 the density of Earth's.

Pluto's small mass means that it cannot be responsible for all the discrepancies in the motions of Uranus and Neptune. This has led some to conjecture that there is another planet even farther out than Pluto.

SATELLITE OF PLUTO

Name	Year Disc.	Distance from Pluto (1,000 miles)	Period of Revolution (days)	Mass (× the Moon's)	Density	Diameter (miles)
Charon	1978	11.68	6.387	0.02 (?)	0.9 (?)	808

A view from Pluto's surface

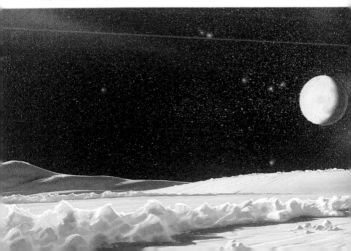

A comet in a 1301 painting by Giotto   ESA

COMETS

Of all the objects in the sky, probably none has caused as much fear as comets. The name means "hairy star" in Greek, for bright comets often appear as diffuse blobs of light with wispy tails. Historically, they have been seen as the finger of an angry god or a sword pointing to Earth. Today, comets are named for their discoverers.

Comets are remnants of material left over from the formation of the solar system. We believe hundreds of billions of them exist in a vast region surrounding the solar system at distances of billions to trillions of miles from the Sun. Once in a while, the orbit of one of these comets is perturbed and it falls toward the inner solar system. If it comes close enough to Earth and the Sun, it may become visible in our skies. It may then return to the depths of space, or the gravity of Jupiter and Saturn may capture it into a permanent orbit among the planets, becoming a periodic comet. If a comet passes very close to the Sun

Comet Halley in 1910
CARNEGIE

(called a "sun-grazer"), it may be broken up, producing two comets, or dispersing completely.

Comets' orbits are mostly highly elliptical, and many are seen only once. A few bright comets are called *periodic* and return predictably in periods ranging from 3.3 years (the shortest known) to about 150 years. The orbits of comets are usually highly inclined to the plane of the ecliptic, and many go around the Sun in a *retrograde* manner.

Strewn along many comets' elliptical orbits are swarms of dust and rock lost by the comet, called meteoroid streams. When the orbit of one of these intersects Earth's orbit, we will see it as a meteor shower each year. We believe most meteors are bits of dispersed comets. The Eta Aquarid and Orionid showers (p. 67) result from meteoroids in Comet Halley's orbit.

ONLY TWO COMETS have been explored closely by spacecraft—Halley and Giacobini-Zinner. The latter was the first comet ever visited by spacecraft, and much was learned that later aided the fleet of Japanese, Soviet, and European space probes that flew by Comet Halley during its passage in 1986.

THE NUCLEUS is the central part of a comet. It is a small object; in the case of Halley, the nucleus is peanut-shaped and about ten miles long and six miles wide. It is a mixture of ice, mostly water ice, with dust and rocks, and it is often called a "dirty snowball." When the comet heats up due to sunlight, the water ice turns to vapor and escapes, jetting through cracks in the surface, leaving a porous interior. The surface is slightly bumpy and is as black as charcoal. Halley's nucleus rotates once every 53 hours.

Comet structure

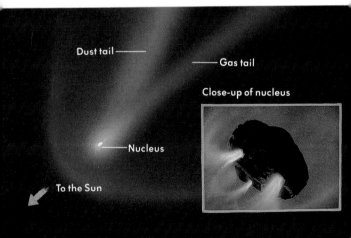

THE COMA surrounds the nucleus. These two parts make up the head of the comet. The coma is a mixture of gases and dust boiled out of the nucleus by the Sun's heat. (Halley boiled away about 25 tons of water per second.) The coma may extend 60,000 miles into space, surrounded by a hydrogen cloud millions of miles across.

COMET TAILS can stretch to hundreds of millions of miles in length, yet they contain very little material. There are really two tails to a comet. The *dust tail* is straight, composed of minute particles freed from the nucleus. It is slightly yellowish in color, from reflected sunlight. The *gas*, or *plasma*, tail is made of ionized gases, often water vapor or carbon monoxide. It is usually curved, and it is usually bluish, the color emitted by these gases. (These colors are faint and can be seen only in photographs.) Since the molecules in the gas tail are electrically charged, they interact with the solar wind and the Sun's magnetic field. Comet tails always point away from the Sun, due to the solar wind, so that when a comet is moving away from the Sun, the comet's tails are actually leading the head.

Every comet loses some of its material with each passage near the Sun. Some of the lost material forms long dust trails ahead of and behind each comet in its orbit. A comet can only make a few hundred such passes near the Sun before it is largely dispersed; this takes thousands to billions of years. During Comet Halley's appearance in 1910, Earth actually passed through the tail of the comet, with no noticeable effect other than a spectacular view of the comet.

Comet tails point away from the Sun

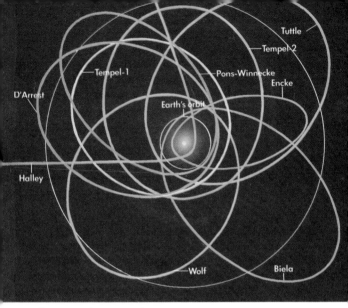

Some periodic comet orbits

OBSERVING COMETS is easy when they are bright. While astronomers may view more than a dozen comets each year through telescopes, only once or twice a decade does a comet appear that is visible to the unaided eye. Most of these are not periodic comets, and cannot be predicted very far in advance. Halley is the only bright periodic comet, and it returns to the inner solar system only once every 76 years (next in 2061).

When far from the Sun, comets are very dim. Predicted comets may be spotted through telescopes when they are beyond the orbit of Jupiter, but new comets are usually not spotted until they are closer to the Sun and thus brighter. Even the brightest comets usually must be within the orbit of Mars to be visible to the unaided eye from Earth. Even

Comet Halley seen by Giotto spaceprobe in 1986   ESA

then, if the comet is on the far side of the Sun, it may be dim and difficult to see from Earth (as was the case with Comet Halley in 1986).

Contrary to popular opinion, comets do not whiz across the sky. Instead, they hang almost motionless against the background of stars, moving slowly among the stars from night to night. Often only the head is seen as a faint, fuzzy ball. The tail, when visible, is a filmy streak of light, brightest near the head.

Only on rare occasions, perhaps a few times a century, such as in 1910 when Earth actually passed through the tail of Comet Halley, will a comet be a truly spectacular object in the sky.

The best instruments for viewing a comet are binoculars because of their wide field and ease of use. Those with telescopes may view several faint comets each year. Be cautioned, though, that comets seen with small instruments will not resemble the professional photographs taken with large telescopes. The excitement of comet viewing is that you are looking at material that is as old as the solar system itself, kept billions of years in a deep-space deep-freeze, holding clues to the origins of ourselves and our planetary neighborhood.

Double star

An imaginary planetary system

PLANETS BEYOND PLUTO?

Astronomers have searched, without success, for major planets beyond Pluto. If any exist, they must be very dark, very small, or very distant—perhaps all of these. They probably resemble Neptune or some of the Jovian satellites.

We have also examined nearby stars for evidence of planetary systems. Even the nearest star system is almost 7,000 times farther from the Sun than Pluto. Any such planets would be too far away to see directly, but we might be able to detect their presence by their gravitational force on the star, making its path through space wavy. There is some evidence that a few planets do exist around nearby stars. The planets we think we detect are huge, many times the mass of Jupiter. There could be other smaller planets in such systems.

Possibly Earth-sized planets orbit these stars in a region where the temperature is suitable for liquid water, and many other conditions are just right. If so, life may have evolved. And "they" may be looking at stars near to them, as we are, wondering if they are alone in the universe.

BIBLIOGRAPHY

For more information, and to keep up with the latest discoveries, consult these and other references:

BOOKS

Beatty, J. Kelly, Brian O'Leary, and Andrew Chaikin, eds., *The New Solar System*, 2nd ed., Sky Publishing Corp., Cambridge, MA, 1982

Chartrand, Mark R., *Skyguide*, Golden Press, New York, NY, 1982

Royal Astronomical Society of Canada, *Observer's Handbook* (yearly), 124 Merton Street, Toronto, Ontario M4S 2Z2 Canada

Zim, Herbert S., Robert H. Baker, and Mark R. Chartrand, *Stars*, rev. ed., Golden Press, New York, NY, 1985

MAGAZINES

Astronomy, AstroMedia Corp., P.O. Box 92788, Milwaukee, WI 53202

Mercury, Astronomical Society of the Pacific, 1290 24th Avenue, San Francisco, CA 94122

Sky and Telescope, Sky Publishing Corp., 49 Bay State Road, Cambridge, MA 02138

Also see occasional articles on astronomy in such magazines as *Science News*, *Scientific American*, *New Scientist*, *Discover*, and *Omni*.

INDEX

Boldface type indicates pages with more extensive information.

PHOTO CREDITS We are indebted to the following institutions and individuals for the photographs used in this book. European Space Agency (ESA): 23 bottom, 150, 155; Lick Observatory Photographs: 74, 77; National Aeronautics and Space Administration (NASA): 20, 21, 22, 23 top, 34, 35, 40, 46, 47, 52, 56, 68, 70 left, 71, 84, 85, 86, 98, 99, 101, 102, 103, 104, 105, 106, 107, 108, 109, 111, 116, 117, 119, 120, 121, 123, 124, 125, 126, 127, 128, 134, 136, 142, 144, 145; The Observatories of the Carnegie Institution of Washington: 19, 70 right, 151; Palomar Observatory Photograph: 146; United States Geological Service (USGS): 63; Jerome Wyckoff: 79